The Immune Memory

How Immunity Learns, Remembers, and Decides

Team Trex

Published by NobleTrex Press

For permissions and other inquiries, write to:

P.O. Box 3132, Framingham, MA 01701, USA

Contents

1 The Body's Invisible Record **7**

1.1 The Archive Without Neurons 7

1.2 The Geometry of Invasion 15

1.3 Hubs, Nodes, and Meeting Points 24

2 Ancient Sensors and New Tricks **33**

2.1 The Hardwired Watchmen 33

2.2 A Storm of Signals 42

2.3 The Epigenetic Ghost 51

3 The Architects of Specificity **61**

3.1 The Intelligence Handoff 61

3.2 Commanders and Executioners 69

3.3 Factories of Precision 77

4 Evolution in Real-Time **85**

4.1 Survival of the Fittest Molecules 85

4.2 The Sentinels That Remain 93

4.3 The Long Wait 100

5 The Mathematics of Survival **109**

5.1 The Calculus of Activation 109

5.2 Logic Gates and the Chemical Internet 117

5.3 The Strategic Value of Noise 126

6 The Delicate Balance of Self **135**

6.1 The Brutal Education of a Lymphocyte 135

6.2 The Peacekeepers 143

6.3 Civil War 151

7 Hacking the System **159**

7.1 The Milkmaid's Paradox 159

7.2 The Dirty Little Secret 167

7.3 The Genetic Instruction Set 175

8 Time and Tenants **185**

8.1 The Trillions Who Tutor Us 185

8.2 The Twilight of the Naïve 194

8.3 The Biography of Exposure 202

9 Decoding the Black Box **211**

9.1 Eavesdropping on the Cellular Monologue 211

9.2 The Signature of Safety 220

9.3 Circuit Boards of Biology 229

10 Engineering the Future **239**

10.1 The Art of Releasing the Brakes 239
10.2 Restoring the Truce 247
10.3 Immunity as Code 255

Introduction

In 2008, a team of researchers led by Eric Yu made a discovery that borders on the miraculous. They were analyzing blood samples from ninety-year-old individuals, survivors of the catastrophic 1918 influenza pandemic that had swept the globe nearly a century prior. These men and women had not encountered that specific H1N1 virus since they were children, yet when the researchers exposed their B cells to the protein structure of the 1918 virus, the cells woke up. They reacted instantly. Within the microscopic architecture of their blood, living sentinels had been patrolling in quiet vigilance for nine decades, waiting for a ghost that never returned-until that moment in the lab.

This is the most enduring memory you possess. It is not the memory of your first kiss, the address of your childhood home, or the Pythagorean theorem. It is a biological memory, written not in neurons and synapses, but in proteins and receptors. It is the reason you are alive to read this sentence.

We are accustomed to thinking of "memory" as a neurological phenomenon, a function of the brain that categorizes our past to help us navigate our future. But long before the first neuron fired, evolution had already solved the problem of remembering. Life exists in a hostile environment, a chaotic soup of pathogens, viruses, and bacteria that are constantly trying to repurpose our machinery for their own reproduction. To survive, multicellular organisms needed a way to distinguish the harmless from the fatal, and more importantly, they needed a way to remember the face of the enemy.

This book is about that second cognitive system: the immune system. We often describe immunity with martial metaphors-it is a defense, a shield, an army. While those comparisons have their utility, they obscure the true genius of the system. The immune system is actually an information-processing machine. It learns. It retains data. It calculates probabilities. And, perhaps most profoundly, it makes decisions.

Every second, your body is making millions of life-or-death choices. A T cell brushes against a pancreatic cell: is this cell healthy self-tissue, or has it been corrupted by a virus? A macrophage encounters a speck of dust in the lung: is this an invading fungal spore that requires a scorching inflammatory response, or is it harmless debris that should be ignored? If the immune system is too passive, infection kills you. If it is too aggressive, it turns its weapons inward, resulting in autoimmune devastation. The margin for error is vanishingly small, yet the system gets it right almost every single time.

How does a collection of single cells, floating in the dark currents of your bloodstream, achieve this collective intelligence? How does a T cell "know" anything?

To answer this, we have to look beyond the simple idea of antibodies fighting germs. We must view the immune system as a sensory organ. Just as your eyes perceive photons to build an image of the visible world, your immune cells perceive molecular shapes to build a map of the chemical world. They read the geometry of proteins. When a virus enters your body, the immune system doesn't see "flu"; it sees a specific arrangement of atoms, a topographical ridge on a protein surface. It snaps a molecular photograph of that shape and files it away.

The sheer scale of this library is staggering. Your body has the potential to generate receptors capable of recognizing nearly every organic shape that exists, or could exist, in the universe. It is a system of infinite improvisation. When a novel pathogen appears-something nature has never created before-your body engages in a process of real-time evolution. Inside your lymph nodes, cells undergo rapid mutation and selection, compressing millions of years of evolutionary trial-and-error into a few days, all to engineer a perfect molecular key for a lock that didn't exist a week ago.

But memory is useless without context. This is where the story of immunity shifts from a mechanical description of locks and keys to a complex narrative of decision-making. For a long time, immunology was dominated by the "self versus non-self" model, a concept solidified by Frank Macfarlane Burnet in the mid-20th century. The idea was simple: the immune system attacks anything foreign and tolerates anything native.

However, we now know this is an oversimplification. We eat "foreign" food daily without going into anaphylactic shock. We host trillions of "foreign" bacteria in our gut microbiome that are essential to our digestion. A mother tolerates the "foreign" DNA of a fetus for nine months. Clearly, the immune system is not just checking ID cards; it is analyzing context. It asks: *Is this dangerous?*

This shift in understanding-from seeing immunity as a simple army to viewing it as a complex, ecologically driven ecosystem-changes everything about how we approach human health. We are realizing that inflammation is not just a symptom of injury, but a language-a chemical syntax that cells use to scream for help or whisper for calm. When this communication breaks down, we see the rise of modern plagues: allergies, asthma, and chronic inflammatory diseases. These are not failures of weakness, but failures of logic-a memory system that has become paranoid, seeing threats where none exist.

Understanding this biological record is no longer an academic exercise; it is a prerequisite for the future of medicine. We are currently living through a golden age of immunology, driven by technologies that allow us to interrogate single cells and map their internal decision trees. We have moved from simply observing the immune system to hacking it.

Vaccination, the oldest immunotherapy, was the first chapter in this story-a way to trick the system into forming a memory without the trauma of the disease. Today, we are writing new chapters. We are engineering T cells to hunt down cancer, releasing the molecular brakes that tumors use to hide in plain sight. We are exploring how to delete specific bad memories to cure autoimmunity, teaching the body to forget its vendetta against the pancreas or the joints. We are even discovering that our "primitive" innate immune system, long thought to be incapable of memory, possesses a form of trained history-a discovery that blurs the lines between our biological distinctness.

This narrative requires us to descend into the microscopic. We will walk through the training grounds of the thymus, a brutal military academy where 98

But we will also zoom out to the mathematical. The immune system is a master of "noisy" computation. It utilizes randomness to hedge its bets against the unknown. It operates on thresholds and feedback loops that look remarkably like the circuits in a computer, yet are built from proteins and signaling gradients. It manages to maintain a distinct definition of "you" despite the fact that your cells are constantly dying and being replaced, and despite the fact that you are teeming with foreign life.

As we dissect these mechanisms, a profound realization emerges: you are not a static entity. You are a historical record. Every cold you fought as a child, every scrape on your knee, every vaccine, and every silent battle your body won while you slept has left a physical mark. Your immune system is a living biography of your interaction with the world. The collection of memory cells in your blood right now is unique in the entire history of the human species; no one else has fought exactly the same battles in the exact same order.

The elegance of this system is terrifying and beautiful. It is capable of unleashing a cytokine storm that can kill you in hours, yet it is precise enough to distinguish between two proteins that differ by a single atom. It is ancient, sharing components with sea sponges and fruit flies, yet it is capable of learning from a virus that evolved yesterday.

So, let us set aside the notion of the body as a fortress. Instead, imagine it as a vast, bustling society of trillions of individuals-blind, deaf, and microscopic-constantly shouting to one another in a chemical language, voting on decisions, trading information, and remembering the past to secure the future. This is the story of how that society works. This is the story of the invisible memory that defines who you are.

Chapter 1

The Body's Invisible Record

We typically imagine memory as a ghost in the neural machine, yet our most vital recollections are written in protein and blood long before they reach our conscious minds. This chapter establishes the physical reality of the immune archive, exploring how the body encodes the history of its microscopic wars into a library of cells ready to protect us from threats we haven't encountered in decades.

1.1 The Archive Without Neurons

A brain remembers by changing itself. A face becomes familiar because billions of synapses, each a tiny adjustable contact, have been nudged into a new pattern. The information is not stored in a single place like a file in a cabinet; it is distributed through altered connections, revised probabilities, strengthened pathways. If you had to point to *where* the memory lives, you would not find a labeled cell marked "Paris" or "first kiss." You would find a network whose wiring has been tuned so that, when the right cue arrives, the whole system falls into a recognisable state.

The immune system also remembers, but it does so without neurons and without synapses. Its archive is not written into a web of connections. It is written into populations. To the immune system, remembering is less like editing a manuscript and more like keeping survivors.

That simple difference explains an everyday mystery: why many second encounters with a virus pass almost unnoticed. A first infection can be a week-long drama; the second can be a footnote, or nothing at all. It is not that the immune system has become "smarter" in a psychological sense. It has become *better stocked.* After the first encounter it keeps, sometimes for decades, a scattered but durable inventory of cells and molecules shaped by that event. When the same invader returns, the response does not begin from scratch. It begins from an advantaged position: more starting material, better tools, and less need to improvise.

To see why, it helps to treat the immune response as a race against time. A virus that doubles every few hours can turn a quiet infection into a crisis before a person feels ill. The immune system must not only recognise the intruder; it must do so quickly enough, and at a scale large enough, to keep the pathogen from reaching its own momentum. The first time the immune system meets a new microbe, recognition is slow for a purely mathematical reason. Among the vast diversity of lymphocytes—the white blood cells that carry highly specific receptors—only a tiny fraction can bind any given microbial "signature." Those cells are rare by design: rarity is the price of breadth. If the body wants to be prepared for an unpredictable world, it must distribute its bets across almost unthinkably many receptor shapes. That means that, for any one pathogen, the winners begin as needles in a haystack.

During a first infection, these rare needles must be found, activated, and multiplied. The multiplication is not gentle. A single activated B cell (the lymphocyte that can become an antibody factory) can produce a lineage that expands into thousands or millions of related cells. A single activated T cell (the lymphocyte that can coordinate immunity or kill infected cells) can do the same. Expansion takes time. Cells must divide, differentiate, move to the right tissues, and switch on the specific tools that make them effective. That delay is why the symptoms of many infections arrive after an incubation period: the early struggle often belongs to the pathogen.

The second time, the system cheats at the arithmetic. The needle is no longer one needle. After the first infection, a fraction of the expanded lineage is not discarded when the battle ends. Instead, it is preserved as a standing force: memory B cells, memory T cells, and long-lived plasma cells (specialised B cells that secrete antibodies continuously). Some of these cells circulate; some take up long-term residence in tissues. Some are poised to proliferate again; others quietly secrete protective molecules year after year. Immunological memory is not just faster because the immune system is "primed." It is faster because the starting population is larger and the cells are already partially prepared.

Place that alongside brain memory and the contrast sharpens. In the nervous system, remembering does not typically mean keeping a large population of "memory neurons" for each experience. The brain cannot afford an ever-growing census of cells dedicated to single events. It uses rewiring and synaptic strength, a compressive strategy that allows immense storage in finite tissue. The immune system, by contrast, can afford to keep a distributed archive of *clones*: families of cells descended from a successful ancestor. Its memory is biological persistence, supported by homeostatic signals that help these cells survive and occasionally divide at low levels, maintaining their numbers even when the original pathogen is long gone.

This does not mean the immune archive is a perfect diary. It is closer to a set of annotated wanted posters. The immune system does not record an enemy in the round; it records the particular molecular features it learned to grab. The archive is thus highly specific, and specificity is both its strength and its weakness. If the returning pathogen is sufficiently similar, the response is swift and often decisive. If it has changed the key features, memory can mislead as well as protect, favouring old targets over new ones. Even at its best, immune memory is not a general-purpose recollection of the past; it is a practical, biased record shaped for future combat.

A historical anecdote makes the logic visceral. In the eighteenth century, long before anyone knew about antibodies or lymphocytes, the English aristocrat Lady Mary Wortley Montagu watched a practice in the Ottoman Empire that was both daring and oddly modern: *variolation*, the deliberate inoculation of a person with material from a mild smallpox case. It was risky, and it was not always mild. Yet it was often safer than waiting for smallpox to arrive uninvited, a disease that could disfigure, blind, or kill. Montagu had survived smallpox herself and had seen its costs. She championed the practice in Britain, arguing from observation rather than mechanism. The gamble only makes sense if the immune system can remember: one controlled encounter, followed by protection from a future catastrophe. When Edward Jenner later developed vaccination using cowpox, the same logic gained a safer method. The invisible record inside the body became a tool that societies could deliberately write.

What, exactly, gets written?

Start with antibodies, because they are the part of immune memory people can picture most easily. Antibodies are proteins secreted by plasma cells into blood and tissues. Each antibody has a binding site whose shape matches a particular molecular feature on a pathogen. When antibodies bind, they can block viruses from entering cells, tag bacteria for ingestion, or recruit other immune mechanisms. After an infection or vaccination, antibody levels in the blood typically rise and then fall. If the story ended there, immunity would be brief.

Yet for many infections, protective antibodies persist for years, sometimes decades, supported by long-lived plasma cells that can reside in niches such as the bone marrow. These cells are less like soldiers waiting for orders and more like factories kept on a maintenance budget, steadily producing a low but meaningful output. They are one physical layer of the archive: an enduring chemical memory.

Another layer is made of memory B cells. Unlike plasma cells, memory B cells do not spend their lives secreting antibodies. They circulate or sit in lymphoid tissues as a reserve, ready to respond if the pathogen returns. Their value is not only speed. During the first encounter, B cells can enter structures called germinal centers, where they mutate their antibody genes and compete for survival based on how well they bind the antigen. The result is a selection process that improves antibody quality: higher affinity, often different antibody classes, and sometimes broader recognition. Memory B cells can preserve the products of that process, keeping a repertoire of improved solutions. On re-exposure, they can rapidly generate new plasma cells, producing a flood of antibodies that is both faster and better targeted than the first wave. Some memory B cells can even return to germinal centers and refine further, a kind of continuing education that allows the archive to be updated.

Memory T cells form a parallel record. Helper T cells support B cell responses and orchestrate immune activity through signalling molecules. Killer (cytotoxic) T cells recognise infected cells and destroy them, limiting viral spread. After an infection, many of the expanded T cell clones die back, but a subset persists as memory. Some memory T cells circulate through blood and lymph nodes, scanning for their antigen. Others are stationed directly in tissues that are common entry points for pathogens: skin, lungs, gut, reproductive tract. These tissue-resident memory T cells act like local specialists. If the same pathogen tries to cross the same border again, these cells can respond on site, often before the pathogen reaches the bloodstream. It is a kind of geographical indexing: part of the archive is stored where it is most likely to be needed.

All of this is "memory" in a biological sense, but it is not a thought, not a belief, not even a sensation. The immune system decides—whether to respond, how strongly, with what tools—yet its decisions are biochemical and cellular, not conscious. The body can "remember" tetanus without knowing what tetanus is. It can mount a rapid antibody response without any awareness. The language of "learning" and "memory" is tempting because it captures the time-dependent improvement, but it can also mislead. The immune system is not a second brain hidden in the chest. It is a set of living processes that change their composition after experience.

The difference matters because it shapes the costs and the risks. A brain can remember a song without spending much energy playing it forever. The immune system, if it wants long-term antibody protection, must pay for factories that keep working. It must allocate space in bone marrow niches, provide survival signals, and accept the long-term presence of cells that might, in rare cases, go wrong. The immune archive is not free. Nor is it unlimited. The body cannot preserve everything at maximal strength, and different pathogens induce different kinds of memory. Some infections, like measles, tend to leave strong, long-lasting protection. Others, like influenza, change their surface features so quickly that last year's record can be only partially useful. Still others elicit weak or short-lived memory for reasons that are not fully about the pathogen and not fully about the host, but about the complex interaction between them.

When the archive works well, the key signature is the difference between a primary and a secondary response. The first encounter is slower, the peak is smaller, and the body often suffers symptoms while the response gathers force. The second encounter rises faster and higher, frequently stopping the pathogen before it announces itself. That is not a guarantee of zero infection: pathogens can still enter, replicate for a time, and even transmit. But the window is narrower, and the damage is usually less. “Usually asymptomatic” does not mean “never infected.” It means the immune system can often keep the battle small enough that the person experiences little or nothing.

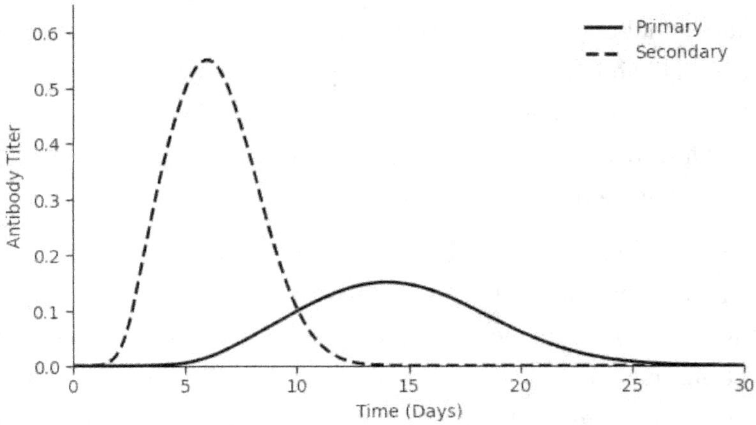

The graph looks almost too simple, but it hides a key truth: speed is not merely a convenience. In infections, speed can be the difference between a mild irritation and a life-threatening cascade. Consider rabies, a virus that is unusually instructive because it moves slowly through nerves toward the brain. Post-exposure vaccination can work even after a bite because there is time to build an immune response before the virus reaches its most dangerous territory. The immune system can be made to write a record quickly enough to stop a lethal disease already in motion. That medical fact is a dramatic demonstration of what immunological memory really is: a capacity for accelerated, improved reaction once the right clones have been established.

Yet the immune archive has to be curated. It must avoid remembering the wrong things. A brain can misremember an event and suffer embarrassment; an immune system that "remembers" the body as foreign can cause chronic disease. The immune system therefore develops under a strict education in tolerance: it must learn, in effect, that many molecular shapes it encounters are *self* and should not be attacked. That education is imperfect, and autoimmunity is one of the prices of a powerful memory system. Even without overt autoimmunity, the system must constantly manage the line between readiness and restraint. Memory cells must be poised to respond quickly, yet they must not react to harmless or familiar cues that resemble a past threat.

There is also a subtler price: a strong archive can bias future responses. If a person encounters a variant of a virus whose surface has changed, memory cells recognising older features may expand rapidly and dominate the response, even if new features would be better targets. This can still be protective—partial recognition is often better than none—but it can also limit how much the immune system invests in building fresh memory against the altered parts. The body, in other words, can become attached to an older file. This is not a moral failing or a cognitive stubbornness; it is a predictable consequence of how cellular persistence and competition work. Cells that are already numerous and easily activated tend to win the early race for resources and attention.

The deeper contrast with the brain now comes into focus. Neural memory is plasticity within a largely stable population: the same neurons can participate in many memories, and the system relies on patterns and weights. Immune memory is plasticity through selection: from a wildly diverse starting pool, particular clones are expanded and then partially preserved. The basic computational move is Darwinian rather than synaptic. Variation exists in advance. Encounter selects. Survivors persist.

That Darwinian logic makes the immune system both astonishingly effective and fundamentally different from anything we introspect. It also means that the immune record is embodied in living things that can die. Immunity can fade if plasma cells are lost or if memory cells decline. Medical treatments that deplete B cells can erase parts of the archive. Aging can change the balance of immune cell populations and the quality of memory. Malnutrition, chronic infections, and immune-suppressing drugs can all affect maintenance. Memory is not a static inscription; it is a managed population ecology inside the body.

Still, when it works, it gives the organism a remarkable advantage: the waste of infection is turned into future protection. The first encounter is costly, sometimes painfully so, but it leaves behind an infrastructure tailored to that specific threat. Antibody factories hum quietly in the background. Memory cells circulate like well-briefed patrols. Tissue-resident sentinels stand at old points of entry. None of these components has any idea what it is doing, and none needs to. The archive is not made of thoughts. It is made of survivors, maintained long after the danger that created them has disappeared, ready to turn a familiar threat into a brief, silent skirmish rather than a full-bodied illness.

1.2 The Geometry of Invasion

A child points at a dog and says “dog” without taking an inventory: not the fur, not the teeth, not the particular curve of the tail. The word lands on the whole. The immune system cannot afford that kind of recognition. It does not see “influenza” or “measles” as a single object any more than a lock sees a key as a story about travel. It sees only shape touching shape, a set of molecular surfaces meeting in three dimensions. In that world, identity is not a name but a fit.

The basic unit of immune recognition is called an *antigen*. In ordinary speech, antigens are often described as "things the immune system reacts to," but that definition hides the key idea: an antigen is usually a molecule or part of a molecule, and what matters is not the whole molecule so much as the particular patch of it that can be grasped. That patch is an *epitope*—a small region on the antigen that a receptor can bind. The language is worth learning because it sharpens how you imagine an infection. A virus is not just an invading particle. To the immune system it is a crowded landscape of epitopes: ridges, loops, grooves, sugars, charges, and textures. Different immune cells can focus on different patches. Even within a single person, the response is a chorus of recognitions, each one narrow, each one literal.

Antibodies make this geometric style of thought easiest to picture. An antibody is a Y-shaped protein with two arms that can clamp onto an epitope. The arms are not rigid pincers; they have flexibility, and their binding depends on a mixture of forces: complementary shape, electrical charge, and small molecular attractions. If you could shrink yourself down and watch an antibody meet its target, you would not see a decisive click so much as a brief dance of near-misses until the right alignment settles into place. When it does, the bond is strong enough to hold through the turbulence of blood flow, but delicate enough that mutations in the pathogen can weaken it.

T cells, which do not bind free-floating antigens in the same way, are no less geometric. They recognise fragments of proteins displayed on the surface of cells by specialised molecules (major histocompatibility complex, often shortened to MHC). This is the immune system's way of forcing an internal sample to the exterior: a cellular bulletin board. A T cell receptor binds the displayed peptide and the MHC together, like a hand assessing not only the object it holds but the glove through which it is held. Here, too, the discriminations are spatial. A change of a single amino acid in a peptide can be the difference between invisibility and alarm.

If immune memory is an archive of survivors, immune recognition is the archive's indexing system. It determines what is filed together and what is treated as unrelated. Two viruses can be close cousins in a taxonomic sense and yet look different to the immune system because their key epitopes differ. Two unrelated microbes can trigger cross-reactions because they share a similar patch. And a virus can remain "the same" in everyday language while drifting antigenically, gradually sanding down the exact surfaces that existing antibodies prefer to grab. This is one reason influenza vaccines are updated: the virus changes the geometry of the most visible parts of its coat, not its entire identity, but enough to evade a significant portion of the existing antibody repertoire.

The idea that immunity depends on molecular fit was once radical because it suggested that the immune system has a near-infinite capacity for specificity. Paul Ehrlich, working at the turn of the twentieth century, imagined that cells carried side chains that could bind toxins, an early conceptual ancestor of receptors. Later, the clonal selection theory, articulated most influentially by Frank Macfarlane Burnet in the 1950s, made the key leap: the body does not sculpt a new receptor to match each invader; it produces a vast diversity of receptors in advance, and infection selects the clones that happen to fit. In this view, a pathogen is not so much "recognised" as it is *matched* by pre-existing variation. The immune system is a library of possible geometries, and disease is the harsh librarian that stamps certain cards "useful" and others "irrelevant."

That library, however, is not composed of whole books. It is composed of excerpts. A single virus can present many epitopes, and the immune system may focus on a subset. Some epitopes are exposed and easy for antibodies to reach; others are hidden in folds or shielded by sugars. Some epitopes are stable because the pathogen cannot easily change them without breaking an essential function; others are disposable decorations, easy to mutate. The immune response, like attention, can be drawn to the loudest features, not necessarily the most strategically important. Vaccines, at their best, try to direct that attention toward parts of the pathogen that are both accessible and constrained.

The language of epitopes also helps explain a small everyday puzzle: why allergies exist. A pollen grain is not a parasite trying to invade tissues. Yet it carries proteins with shapes that, in some people, elicit antibody responses of a particular kind (IgE) that recruit mast cells and trigger swelling, itch, and sometimes asthma. The immune system is exquisitely tuned for recognizing shape, but it must also decide what a shape *means*. Without context, a shape is just a shape. The same geometric recognition that protects against parasites can misfire against harmless environmental molecules. This is not simply a matter of an overexcited immune system. It is a consequence of how much of immunity depends on associative signals: injury, infection, and the chemical "accent" of inflammation.

A concrete case makes the geometry vivid: the way certain antibodies can *neutralise* viruses by blocking the specific region the virus uses to attach to our cells. Imagine a virus as a burr that must hook onto a particular fabric to enter. If an antibody binds over the hook, the virus becomes a harmless sphere, unable to get traction. That is neutralisation in simple terms. The extraordinary success of some vaccines comes from inducing large quantities of such antibodies against critical viral attachment sites. When those sites change shape, neutralisation can weaken even if the immune system still recognises other parts of the virus. The infection may then be less severe rather than fully prevented. This is one reason the word "breakthrough infection" can be misleading: the immune system may have *partly* succeeded by transforming a potential pneumonia into a mild cold.

To understand how the immune system finds these geometric targets in the first place, you have to imagine how antigens travel inside the body. A cut on the finger, a bite on the ankle, a virus inhaled into the lungs: these are local events. Yet the adaptive immune response, with its rare specific lymphocytes, cannot rely on a few scattered cells in the tissue happening to be nearby. It needs meeting points. Much of the body's immune geometry is therefore not only molecular but anatomical. Antigens do not merely exist; they are routed.

The lymphatic system is one of the great pieces of biological infrastructure that most people never think about until a lymph node swells. Fluid from tissues drains into lymphatic vessels and is carried to lymph nodes, bean-shaped organs that function like filtration hubs. The immune system uses this drainage as a delivery service. Antigen from a peripheral tissue is swept into a node where large numbers of lymphocytes are constantly circulating. This does not guarantee that a matching cell will be present, but it increases the odds enough to make the system workable.

Within a lymph node, geometry becomes choreography. Different regions are specialised for different cell types, and chemical signals act as invisible signposts. B cells tend to occupy follicles; T cells tend to occupy a paracortical zone. Dendritic cells, the professional antigen-presenters that pick up material in tissues and carry it to lymph nodes, navigate into T cell areas. A handful of chemokines—small guiding proteins—organise this spatial separation and controlled mingling. The result is not random mingling but structured encounter: the right kinds of cells are brought into the same neighbourhoods with the right kind of antigen display.

Even the node's outer edge is an active interface. Lymph arrives through afferent vessels and empties into a space under the capsule called the subcapsular sinus. This perimeter is not simply plumbing; it is a surveillance zone. Specialised macrophages positioned there can capture viral particles and immune complexes from the incoming lymph within minutes. In effect, the node has bouncers at the door. They intercept what comes in, reduce the chance that infectious particles flood directly onward, and present captured material in ways that make it accessible to B cells in nearby follicles.

This peripheral capture is also an information-routing system. Many antigens arrive in forms that would be hard for a rare B cell to find in the vastness of a node. By holding antigen at defined sites and relaying it inward, the lymph node creates a stable display that increases the probability that a matching B cell will encounter its epitope. There is an elegance here: the immune system does not merely rely on molecular affinity; it manufactures opportunity.

The relay can be surprisingly indirect. A B cell whose receptor does *not* match a particular antigen can still participate in transport by physically carrying immune complexes deeper into follicles, handing them off along a chain. This is not altruism; it is a side effect of how cells move and how they handle material attached to their surfaces. The consequence, however, is profound: the quality of antibody memory is influenced not only by which epitopes exist, but by how efficiently those epitopes are delivered into the specialised microenvironments where B cells compete and refine. Germinal centers, where affinity maturation occurs, are not merely places where B cells mutate; they are arenas supplied by a logistical network.

The story of immune recognition, then, is not simply lock-and-key at the molecular scale. It is also architecture and traffic engineering at the tissue scale. A person can inhale a virus into the airway, and yet the decisive encounters that shape long-term antibody memory may occur in a lymph node some distance away, where antigen has been sampled, concentrated, and displayed in a way that turns a diffuse invasion into a graspable problem.

A historical episode helps show how far this geometric understanding can go in practical terms. In the 1980s, structural biologists began resolving the three-dimensional shapes of proteins with increasing clarity. By the 1990s and 2000s, the structures of viral proteins bound to antibodies became a recurring theme: you could see, almost literally, where protection attached. This was not mere scientific satisfaction. It changed how vaccines and antibody therapies could be designed. Instead of asking only whether a vaccine induces antibodies, researchers could ask whether it induces antibodies against the *right* surfaces: the conserved, functionally constrained epitopes that viruses struggle to mutate away. The immune system always worked in this geometric manner; what changed was our ability to see the contours it was responding to.

That ability also clarified why some immune responses fail. A bacterium can cloak itself in a capsule of sugar molecules that are poor at provoking strong T cell help, dulling the quality of the antibody response. A virus can hide vulnerable epitopes behind flexible loops or a forest of sugars, presenting decoy surfaces that attract antibodies that bind well but do little. Some pathogens exploit a cruel trick: they display epitopes that resemble our own molecules closely enough to make the immune system cautious, pulling the response toward tolerance and away from attack. The immune system's commitment to avoiding self-destruction becomes an exploitable feature.

Even within one infection, the immune system's view can change over time. Early in an infection, the most accessible epitopes dominate because they are the first to be seen and the easiest to bind. As antibodies accumulate, they can mask certain surfaces and shift what is available. As viruses mutate under pressure, some epitopes vanish and others become newly exposed. The immune response is therefore a moving conversation between recognition and escape. Memory preserves a record of that conversation, but it preserves it unevenly, with emphasis on what was successful and prominent.

This brings back the subtlety that sits behind the everyday claim that prior infection or vaccination "trains" the immune system. Training is not simply a matter of speed. It is a matter of *which shapes become familiar*. A person's immune history creates a preference profile: some epitopes are strongly represented in memory, with abundant high-affinity antibodies and readily activated memory B cells; other epitopes are barely noticed, even if they exist on the same pathogen. When a related pathogen appears, the immune system may respond most vigorously to what looks familiar, even if the most dangerous parts have changed. That is not irrational; it is the consequence of an archive built by selection. In evolution, using a previously successful strategy quickly is often wise. The cost is that the system can be conservative when novelty would be valuable.

There is a way to feel the geometry in your own body without microscopes or jargon. Consider why skin tests for allergies work. A tiny amount of an allergen is introduced just under the skin. If a person has IgE antibodies specific for an epitope on that allergen, those antibodies are already bound to mast cells like tripwires. The allergen cross-links them, and the mast cells release histamine and other mediators, producing a local wheal-and-flare reaction. The test is not looking for a general sensitivity; it is probing whether a specific geometric recognition already exists and is wired into a particular effector pathway. The swelling is a visible shadow cast by molecular fit.

Or consider blood typing. The ABO blood groups differ by small sugar structures on the surface of red blood cells. These are epitopes. People naturally carry antibodies against the ABO epitopes they do not have, likely formed in response to similar carbohydrate structures in gut microbes. A mismatched transfusion can trigger rapid destruction of donor red cells because the immune system is exquisitely literal about these surface geometries. Here the immune system is not reacting to an infection at all, but to the presence of an unfamiliar molecular pattern on cells that, in every other respect, look human.

What keeps the system from reacting to everything unfamiliar? Part of the answer lies in context: the chemical environment of inflammation, the presence of danger signals, the way antigen is presented by antigen-presenting cells. Another part lies in thresholds. Binding alone is often not enough. Many immune cells require multiple signals: receptor engagement plus co-stimulation, or antigen plus cytokines that indicate infection. This is one of the immune system's most consequential "decisions," and it is made without consciousness, through molecular logic gates. The geometry of an epitope can open a door, but the tissue context determines whether the cell walks through it or waits.

The immune system's sensitivity to shape is therefore not an abstract curiosity. It is the reason vaccines can work with a fragment of a pathogen rather than the whole organism. It is the reason variants matter even when a virus remains recognisable to us. It is the reason that two people can have different outcomes from the same exposure: their immune archives emphasise different epitopes, and their systems have different histories of what shapes have been stamped as urgent.

Invasion, from this perspective, is not a single event but a continual presentation of surfaces. The pathogen sheds proteins, displays them, mutates them, hides them, and is physically carried into the anatomical spaces where immune cells can inspect them. The body responds by sampling, concentrating, and comparing those surfaces against a vast repertoire of possible fits. In the end, the immune system does not need to picture the enemy as a whole. It only needs to grasp enough of the right edges, with enough strength and speed, to turn replication into a dead end.

1.3 Hubs, Nodes, and Meeting Points

A single lymphocyte is a small, stubborn fact: it can recognise one molecular shape, and it is rare. If the body kept only a few million distinct receptors on offer, recognition would be easy but brittle. Instead it maintains an astronomical variety of receptors across its B cells and T cells, each variety represented by only a handful of cells at any given moment. That strategy is brilliant against the unknown, but it creates a logistical problem that is almost comic in its cruelty. Somewhere in the body, perhaps in the skin of a finger or the lining of the lung, a virus begins to multiply. The lymphocyte that can recognise it may be drifting through the bloodstream on the other side of the body, or wandering inside a different lymph node entirely, or pausing in the spleen as if waiting for a train. How do two strangers in a vast organism find each other fast enough to matter?

The immune system's answer is not to make every tissue a full-service battlefield. That would be wasteful and dangerous. The answer is to build hubs—specialised meeting points—and to force information through them. The lymphatic system, which many people think of as a passive drainage network, is better understood as a distribution grid for immune surveillance. It gathers what happens in tissues and brings it to places designed for searching. The lymph node is not merely a swollen annoyance in the neck during a cold; it is an engineered market where rare buyers and rare sellers are compelled to cross paths.

A lymph node is fed by lymphatic vessels that collect fluid from nearby tissues. This fluid, lymph, is not a simple filtrate. It carries proteins, debris, fragments of microbes, and sometimes intact pathogens. It also carries immune cells that have been active in the tissue. When lymph arrives at a node, it first passes into a space under the capsule, the subcapsular sinus, like water entering a harbour. Along this margin sit specialised macrophages, large cells that act as both scavengers and sentries. They can snatch particles out of the flow with startling speed. In the case of viruses, this capture is not just cleanup. It is the first act of presentation: a way of holding antigen in a stable location where B cells can sample it.

This edge-of-node interception changes the odds. In an open tissue, antigen may be diluted and fleeting, and the few relevant B cells may never touch it. In the lymph node, antigen can be retained and displayed at a boundary that B cells patrol. Even more striking, antigen can be relayed inward. Immune complexes—antigen bound by antibody or complement—can be held at the node's periphery and then passed, directly or indirectly, toward follicles where B cells cluster. Some of the couriers in this relay are B cells that are not specific for that antigen at all. They act as transport, not as recognisers. It is an arrangement that feels almost bureaucratic: the parcel is moved along a chain of hands until it reaches the desk where the specialist sits.

T cells face a different version of the same problem because they are activated not by free antigen but by antigen presented on the surface of antigen-presenting cells, especially dendritic cells. Dendritic cells in tissues behave like professional collectors. They extend processes into the environment, take up material, and, when they detect the molecular cues of infection or injury, they travel. Their destination is usually the draining lymph node, the particular node whose lymphatic vessels receive fluid from that tissue. A virus in the skin of the forearm does not, at first, become a whole-body message. It becomes a message delivered to a specific address.

The interior of a lymph node is a map, not a bowl of soup. Immune cells do not wander in random three-dimensional chaos. They move along tracks and through zones that are chemically signposted. B cells tend to reside in follicles, where a chemokine called CXCL13 helps position them. T cells and many dendritic cells tend to occupy the paracortex, guided by other chemokines such as CCL21 (and its close relative CCL19), which attract cells carrying the receptor CCR7. If this sounds like a minor technicality, consider what it achieves: it keeps B cells near other B cells and antigen displays suited to them, and it keeps T cells near dendritic cells presenting peptides on MHC molecules. The node is organised to maximise productive collisions and to minimise distracting ones.

The node also has entry gates. Lymphocytes circulating in blood do not simply leak into nodes like water through a sponge. They enter through specialised blood vessels called high endothelial venules (HEVs). These vessels act like customs checkpoints. Lymphocytes roll, adhere, and then squeeze through into the node, guided by chemokines displayed on the vessel surface. The chemokines are not merely floating signals; they are presented in a structured way, held in place so that cells passing by can sense a stable gradient at precisely the moment they need to decide whether to exit the bloodstream. The details of this presentation matter: the vessel wall is not only an inert pipe but a carefully prepared surface that can display chemokines effectively.

Once inside, lymphocytes do what seems, at first glance, inefficient. They move. They roam within their zones, scanning cell surfaces, making brief contacts, then moving on. A T cell may contact many dendritic cells in a short time, sampling the peptides on display, and only rarely encountering the one that bears its matching fragment. This scanning behaviour can look like aimless wandering, but it is a search algorithm implemented by cells. The node's architecture makes the search plausible: dendritic cells spread their processes through the T cell zone like trees in a dense orchard, creating a lattice of possible contacts. The T cell need not find a single target cell in a void; it must find one among a well-distributed network.

The stromal skeleton of the lymph node is often overlooked because it is not made of the glamorous immune cell types. Yet it is essential. Fibroblastic reticular cells (FRCs) build a meshwork that provides both structure and instruction. They form conduits that can guide the movement of small molecules and help organise where cells travel. They also produce survival factors and chemokines that maintain the node's zones and keep the right cells in the right neighbourhoods. A lymph node is thus not simply a bag of leukocytes. It is a living organ whose "furniture" tells its occupants where to sit and how long they can stay.

These logistics explain an everyday clinical fact: why lymph nodes swell during infection. Swelling is not only inflammation in the loose sense of the word. It reflects increased traffic and proliferation. More lymphocytes enter through HEVs. Dendritic cells arrive from tissue carrying antigen. Activated lymphocytes divide, increasing local cell numbers dramatically. The node is temporarily transformed from a calm transit station into a crowded operations room. People often notice this in the neck during a sore throat because the nodes draining the throat are close to the skin and easy to feel. The swelling is a palpable sign of an invisible meeting that has gone well: the right rare cells have found their target and begun to expand.

A vivid historical example of nodes as meeting points came from an unexpected direction: the early era of organ transplantation. When surgeons began transplanting kidneys with increasing success in the mid-twentieth century, one of the great obstacles was rejection. The immune system, trained to distinguish self from non-self, treated the transplanted organ as an intruder. Researchers learned, sometimes painfully, that rejection could be accelerated or delayed depending on where antigens from the graft were presented and how immune cells were activated. Lymph nodes draining the transplanted tissue became key sites of immune priming. The practical lesson was anatomical: immunity is not only about what is foreign, but about where the evidence of foreignness is carried and displayed. Modern immunosuppressive strategies, and the careful matching of donor and recipient tissue types, are ways of intervening in this geography of decision-making.

The same principle underlies vaccination in a more benign form. The site of injection is not chosen only for convenience. Injecting into muscle or skin delivers antigen into tissues rich in dendritic cells and lymphatic drainage. The antigen does not need to diffuse everywhere; it needs to reach a draining lymph node efficiently. The swelling or soreness near a vaccine site can be a minor echo of this traffic: local inflammation increases antigen uptake and the movement of antigen-presenting cells, and the nearby node becomes a hub of activation. Even when a vaccine uses a purified protein or a genetic instruction that cells translate into protein, its success depends on the same logistics: antigen must be delivered to the right place, and immune cells must be guided into productive encounters.

The lymph node is not the only hub. The spleen performs a similar function for blood-borne antigens, filtering blood rather than lymph. Mucosal tissues have specialised lymphoid structures that sample the gut and airways. Yet the general strategy repeats: concentrate information, concentrate searchers, and enforce contact by routing traffic through narrow points.

It is tempting to imagine this as a benevolent design, but it is more accurate to call it a compromise forced by scale. The body is large. The immune repertoire is vast. The relevant cells are rare. Search must be centralised, and centralisation comes with costs. Hubs can be overwhelmed. Some pathogens exploit them, using the very mechanisms of trafficking to spread. Others avoid them by remaining local or by suppressing the signals that recruit dendritic cells and lymphocytes. Still, the hub strategy is so effective that it can often reduce a systemic threat to a local skirmish.

This spatial logic also leaves an imprint on immune memory. The immune system does not merely remember *what* it has seen; it also remembers, in a sense, *where* it has been needed. Tissue-resident memory T cells are a striking example. After certain infections, a population of T cells remains lodged in the tissue where the pathogen entered, such as the skin or the lung. They do not recirculate in the same way as other memory cells. They are stationed. From a logistical standpoint, this is an answer to the same problem that created lymph nodes: how to ensure that rare recognisers are present at the right time. If the pathogen is likely to return to the same portal, posting guards there reduces dependence on distant hubs. It is an investment in local speed.

At the scale of a single node, the idea of "decision" becomes concrete. A naive T cell entering a lymph node through an HEV is not committed to anything. It is a cautious, untested cell. It will scan. If it finds nothing, it leaves and continues its circulation, perhaps exiting via efferent lymphatic vessels and returning to blood. If it encounters its antigen on a dendritic cell, and if the context provides the right co-stimulatory signals and cytokines, it changes state. It becomes activated, begins to divide, and differentiates into effector cells suited to the particular kind of threat. The decision is not a moment of contemplation; it is a shift in gene expression, metabolism, movement patterns, and survival requirements. The node's architecture makes the moment possible. Without the hub, the decision would rarely occur because the relevant meeting would rarely happen.

A similar story holds for B cells. A B cell that binds antigen strongly enough can internalise it, process it, and present fragments to helper T cells. Those helper T cells, if appropriately activated, can provide support that allows the B cell to proliferate and enter germinal centers, where antibody genes mutate and are selected. Here, too, the meeting point is crucial. B cells and T cells are not naturally mixed; they are zoned. Their collaboration requires a controlled border crossing within the lymph node, guided by changing chemokine receptor expression that draws activated cells toward each other. The immune system is constantly negotiating who should meet whom and under what conditions.

The physicality of these meetings explains why immunology is full of "microenvironments." It is not a decorative term. The fate of a lymphocyte depends on which cytokines bathe it, which stromal cells support it, which neighbouring cells touch it, and how long those contacts last. A lymph node is a mosaic of such microenvironments, each one tuned to push cells toward particular outcomes. Memory, in this view, is not only a record kept by surviving clones. It is also a record shaped by the places those clones passed through. The archive is written in a specific handwriting because it was written at a specific desk.

One can feel a faint emotional analogue to this in the difference between knowing something in the abstract and learning it in a particular room with particular people. The immune system's version is biochemical rather than psychological, but the logic is comparable: context shapes the kind of memory that is laid down. An antigen encountered with the alarms of infection produces a different quality of memory from the same antigen encountered in a quiet context, such as food proteins in the gut, where tolerance is often the wiser response. The hubs and their microenvironments help the body avoid overreacting to the constant flood of harmless molecules while still responding fiercely to genuine invasion.

All of this infrastructure exists to solve a search problem under severe constraints. The immune system cannot call a meeting of every lymphocyte whenever a pathogen appears; that would be like evacuating a city every time someone coughs. Instead it creates places where the right meetings can happen quickly, and it runs a continuous circulation of cells through those places, like commuters passing through a central station. The elegance is not in the metaphor but in the economics: a limited number of hubs can coordinate surveillance across a vast body.

When a pathogen enters through the skin, it is not met by a single all-knowing defender. It is met by a series of filters and rendezvous points. Some of these are local and immediate: tissue macrophages and resident T cells that can respond within minutes or hours. Some are central and deliberate: dendritic cells carrying evidence to a node, lymphocytes entering through HEVs, antigen displayed at the subcapsular sinus, B cells sampling follicles, T cells scanning dendritic networks. The outcome depends on whether the right rare cells arrive in time and receive the right combination of signals to commit. When they do, the multiplication that follows is dramatic enough to change the balance of the infection, and the survivors of that multiplication become part of the long-term archive.

A swollen node, then, is not merely a sign that the immune system is "active." It is the visible price of coordination. Inside that tenderness is a crowd of cells making and breaking contacts, comparing molecular shapes, exchanging signals, competing for survival cues, and committing to fates that will last long after the immediate infection is forgotten. The body's invisible record is written not in a single organ but across a network of hubs, each one turning the vastness of the organism into a solvable problem of meeting, recognition, and response.

Chapter 2

Ancient Sensors and New Tricks

For decades, the innate immune system was dismissed as a crude, forgetful blunt instrument that merely held the line until the 'real' specialists arrived. We now know that these ancient defenders possess a sophisticated ability to interpret danger and, surprisingly, rewire their own DNA to remember past battles-a phenomenon that challenges our fundamental definition of biological memory.

2.1 The Hardwired Watchmen

A splinter under the skin is a small injury, but it is also a test. The body has only minutes to decide whether the splinter is sterile wood that simply needs clearing, or a vehicle for microbes that should trigger a full alarm. That decision cannot wait for careful deliberation. It begins with a set of molecular sentries already stationed in tissues: cells that are not searching for a particular named enemy, but for the kinds of signatures that *usually* accompany danger. Immunology calls these signatures *patterns*, and the sensors that detect them *pattern recognition receptors* (PRRs).

PRRs are the hardwired watchmen of immunity. The phrase *hardwired* matters. These receptors are encoded in the genome in a relatively fixed form, passed from parent to child. A newborn does not need to *learn* how to recognize bacterial cell-wall fragments or viral genetic material in the wrong cellular compartment; those sensitivities arrive as part of the species' inherited equipment. They are not perfect, but they are fast. When they are triggered, the consequence is not merely a vague flare of inflammation. The cell turns on particular genetic programs: sets of genes that recruit help, change the local environment, and shape what kinds of immune responses are likely to follow.

The contrast with adaptive immunity is striking. T cells and B cells rely on antigen receptors that are not simply inherited. During development, these cells assemble their receptors by rearranging gene segments, generating enormous diversity. The advantage is precision: the adaptive system can, in principle, produce receptors for almost any molecular shape a pathogen can present, even if evolution has never encountered it before. The cost is time and risk. Time, because a rare T or B cell with a useful receptor has to be found, activated, and expanded into an army; risk, because generating diversity can also generate self-reactive receptors that must be controlled.

Innate immunity is often described as a first line and adaptive immunity as a second line, but that can make innate responses sound like a crude placeholder. A better picture is an emergency dispatch system paired with an investigative bureau. Dispatch must respond quickly to incomplete information: smoke, broken glass, an alarm. The investigators arrive later and can identify the culprit with exquisite specificity. The success of the whole enterprise depends on the interface: the quality of the early call, the appropriateness of the response, and the ability to avoid catastrophic overreaction.

The speed difference is not subtle. Many innate responses begin within minutes. A macrophage (a large tissue-dwelling immune cell) that senses bacterial components can start secreting cytokines—

small protein signals that influence nearby cells—in well under an hour. Blood vessels in the area become more permeable, allowing plasma proteins and immune cells to enter tissue. Neutrophils, the body's rapid-response phagocytes, can be recruited quickly and arrive in large numbers. This early choreography can contain an infection before it becomes visible as illness. Adaptive responses, by contrast, typically require days to reach full strength. Even when memory cells are present, the system still needs time to proliferate, differentiate, and deploy.

This difference in timing is partly a matter of genetics. A macrophage does not need to invent a receptor for a bacterial cell wall; it expresses receptors that evolution has already tuned to recognize broad categories of microbial molecules that are difficult for microbes to change. Many bacteria have cell walls that include lipopolysaccharide (LPS) or peptidoglycan; many viruses produce double-stranded RNA at some point in their replication. These features are not unique to a particular strain, but they are informative: they reliably indicate a kind of intruder. PRRs exploit that reliability.

A central set of PRRs in humans are the *Toll-like receptors* (TLRs). Their story contains a useful historical twist. The name *Toll* comes from work in the fruit fly, *Drosophila*. In the 1980s and 1990s, researchers studying fly development identified a gene called *Toll*. Later, it became clear that the Toll pathway also played a role in the fly's defense against infection. This was not, at first, the mainstream way immunologists thought about host defense in humans. For much of the twentieth century, immunological glory belonged to antibodies, vaccines, and the emerging understanding of T cells: the adaptive system, with its elegant specificity. Innate immunity was sometimes treated as background noise.

A key conceptual shift came from the insistence that the immune system needs more than antigen recognition to act appropriately.

Charles Janeway argued in 1989 that the immune system must be able to sense infection-associated context: not just *what* molecule is present, but whether it arrives with the hallmarks of microbial danger. When mammalian Toll-like receptors were identified and linked to microbial detection in the late 1990s, the field gained a molecular footing for this argument. It became possible to say, concretely, that certain receptors respond to conserved microbial features and trigger the expression of co-stimulatory molecules and cytokines needed to activate T cells effectively.

To appreciate the logic, it helps to be explicit about what PRRs are *not*. They are not like antibodies that bind a unique epitope (a small portion of an antigen) with tailored precision. A PRR is more like a smoke detector than a face-recognition camera. It does not tell you which exact brand of cigarette is burning; it tells you that combustion is happening. A smoke detector placed in a kitchen is useful because it is calibrated to an event that tends to be dangerous. PRRs are calibrated to molecular events that tend to indicate microbial invasion or tissue damage: foreign nucleic acids in unexpected places, cell-wall fragments, changes in ion flux, disrupted membranes.

The cellular geography of these sensors is part of their intelligence. Some TLRs sit on the cell surface, poised to sample the outside world. Others are tucked into endosomes, the internal compartments that form when a cell engulfs material. This matters because nucleic acids are everywhere in biology, including in our own cells. Detecting viral RNA is helpful; detecting the cell's own RNA would be disastrous. Placing nucleic-acid-sensing receptors in endosomes helps reduce accidental activation. Viral genomes often enter through endocytosis and are uncoated in these compartments; the receptor sees viral RNA where it expects danger.

There are also PRRs in the cytosol, the fluid interior of the cell, where foreign nucleic acids are particularly suspicious. One widely discussed example is the cGAS–STING pathway, which detects double-stranded DNA in the cytosol. Under normal conditions, DNA is kept in the nucleus and mitochondria. Cytosolic DNA can indicate a DNA virus, but it can also indicate cellular distress: damaged mitochondria leaking DNA, or errors in cell division producing micronuclei that rupture. The benefit of this sensor is obvious: it catches stealthy infections that avoid endosomes. The danger is equally obvious: mistakes in compartmentalization can turn the pathway against the host, contributing to chronic inflammation and autoimmunity.

This is the tightrope walk of innate immunity. The system is built to be quick, and quick systems are often blunt instruments. Yet innate responses are not merely blunt; they are *programmable* in a constrained way. Different PRRs feed into different signaling pathways and transcription factors. The most famous of these transcription factors, because it appears again and again, is NF-κB, a master switch for inflammatory genes. Others include IRFs (interferon regulatory factors), important for antiviral type I interferon responses, and MAP kinase pathways that help sculpt the pattern of gene expression. The details can become technical, but the overarching idea is simple: the receptor does not just say *danger*; it biases the cell toward a response that fits a category of danger.

A bacterial invasion in tissue often triggers strong inflammatory cytokines such as TNF and IL-1, chemokines that recruit neutrophils, and activation of complement and other antimicrobial effectors. Viral nucleic acids often trigger type I interferons, which place nearby cells into an antiviral state, reduce viral replication, and enhance antigen presentation. These are not separate worlds; bacteria can trigger interferons and viruses can trigger inflammation. But there is a meaningful difference in emphasis, and that emphasis is partly determined by which sensors are triggered and where.

The speed of innate immunity does not come only from receptors; it comes from the kinds of cells that carry them and the actions those cells can take immediately. Neutrophils are produced in vast numbers and circulate ready for rapid deployment. Macrophages and dendritic cells reside in tissues as sentinels, already positioned at likely entry points: the skin, lungs, gut, and other barrier sites. Many innate defenses are pre-made or quickly mobilized: antimicrobial peptides that can punch holes in microbial membranes, enzymes that degrade bacterial walls, reactive oxygen species that damage pathogens, and the complement system in blood plasma that can opsonize (coat) microbes and help kill them.

By contrast, adaptive immunity is a system built around scarcity and amplification. A T cell with a receptor specific to a new pathogen might be one in a million. Finding it is like looking for a particular key on a key ring the size of a city. The body solves the problem by having antigen-presenting cells, especially dendritic cells, gather samples and present peptides (small fragments of proteins) on MHC molecules to T cells in lymph nodes. When the right T cell finally recognizes its peptide–MHC combination and receives appropriate co-stimulation and cytokine signals, it proliferates explosively. The delay is the price paid for specificity and breadth.

The relationship between the two systems is often misunderstood as a simple relay: innate first, adaptive second. In reality, the innate system is also a filter and a teacher. It influences whether an adaptive response is initiated at all, and what sort of response it becomes. This is not a mystical intelligence. It is the ordinary logic of signaling molecules and cellular differentiation. Cytokines present during T cell activation influence whether helper T cells become more suited to dealing with intracellular pathogens, extracellular bacteria, or parasites; they influence the balance between inflammation and tolerance. The early signals shape the later architecture.

A concrete example makes the trade-off vivid. Consider a cut on a finger contaminated with soil. The microbial world of soil is huge; there is no time to identify each organism with molecular precision before doing anything. Innate sensors detect common bacterial and fungal motifs, leading to a local inflammatory response. Neutrophils arrive and phagocytose microbes. Complement proteins bind microbial surfaces. The redness and warmth of the area reflect increased blood flow and vessel permeability. These changes are not collateral damage for its own sake; they are part of a strategy to deliver resources and cells to a site where speed matters.

If some microbes persist, dendritic cells that have sampled them will carry antigens to a lymph node. There, adaptive responses begin. The system does not start from nothing; it starts from an environment already shaped by the innate decision. If the early signals suggest bacteria that live outside cells, the ensuing helper T cell responses tend to support antibody production and neutrophil recruitment. If the early signals suggest a virus living inside cells, the response tends to favor cytotoxic T cells and interferon-driven defenses. In practice, many infections are mixed or ambiguous, and the immune system blends strategies. Still, the early bias matters.

One can also see the complementarity in vaccine design. Some vaccines work best when they include *adjuvants*, substances that stimulate innate sensors and make the immune system take the antigen seriously. The antigen provides the target for specificity; the adjuvant provides the context. The idea is old in practice and newer in molecular clarity. Once TLRs and other PRRs were characterized, adjuvants could be understood not simply as irritants but as tools that engage particular pathways and therefore shape the quality of immune memory.

The innate system's hardwiring does not mean it is rigid in every respect. It means its recognition rules are evolutionarily pre-specified: it will respond to certain molecular patterns without needing prior exposure. But even within that constraint, there is room for calibration. Cells differ in which receptors they express. The same receptor can produce different outputs depending on the cell type, its metabolic state, and the broader cytokine environment. A macrophage in the lung is not the same as a macrophage in the liver. The point is not that innate immunity is simple; it is that its recognition layer is encoded at the species level rather than generated anew in each individual.

To hold the contrast in mind, it helps to visualize two curves: one for speed, one for specificity. Innate immunity rises quickly, often within minutes to hours, but it aims at broad categories. Adaptive immunity lags, often days, but it can lock onto a particular strain with exquisite discrimination.

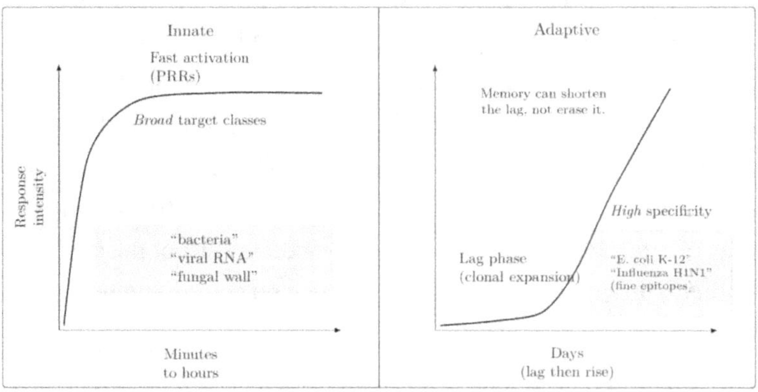

The diagram hides a final nuance. The innate system is fast, but speed alone does not explain why it can be so decisive. A deeper reason is that it is allowed to act on partial information. If a macrophage waits to identify the precise strain of bacterium before recruiting neutrophils, it may lose the race. Acting on category-level information is often sufficient in the early phase. This is why evolution has favored receptors that detect features that are *functionally constrained* for microbes. A virus can mutate rapidly, but it still needs to replicate nucleic acids; a bacterium can swap surface proteins, but it still needs a cell wall and ribosomes. PRRs aim for the parts of microbial life that are hard to abandon.

At the same time, acting on partial information explains why innate responses sometimes harm us. The same inflammatory mediators that help control infection can damage tissues if they are too intense or too prolonged. The body therefore builds brakes into the system: anti-inflammatory cytokines, receptor desensitization, cell death programs, and regulatory circuits that try to keep the alarm proportional. These brakes are not always sufficient, especially when pathogens exploit the system or when the signals of damage become self-sustaining.

The adaptive system is often portrayed as the *smart* part because it is specific, but specificity is not the only kind of intelligence. There is intelligence in being able to respond at all when there is no time to learn. A smoke detector cannot tell you who started the fire, but it can wake you up in time to survive. The hardwired watchmen do something similar at the molecular scale: they translate the first hints of microbial intrusion into an immediate shift of the tissue environment, buying time, recruiting forces, and setting the terms under which precision can later be achieved.

A finger cut that becomes infected is unpleasant, but it usually remains local. The reason is not that the microbes are weak; it is that the early response changes the battlefield. Blood flow increases. Cells arrive that can engulf and kill. The tissue becomes chemically hostile to microbes. The system does not need to identify the invader's full molecular biography to do these things. It needs only to recognize that the situation resembles the kinds of situations in which delay is costly.

When it works well, the innate response can resolve the problem so early that the adaptive system barely needs to mobilize. When the problem is larger, the innate response still matters, because it is the first draft of the immune system's decision: not a final verdict, but a rapid, genetically encoded judgment that turns threat into signal and signal into action.

2.2 A Storm of Signals

Redness, heat, swelling, pain: the classic signs of inflammation are so familiar that they can feel self-explanatory, as if the body is simply *reacting* and, in the process, getting messy. That picture is tempting because inflammation often *does* feel like collateral damage. A throat burns, a joint aches, lungs tighten, skin erupts. From the inside, it is easy to interpret inflammation as the immune system losing its temper.

Yet inflammation is less like temper and more like communication. It is the body's fastest way of turning a local event—a splinter, a virus-infected cell, a torn muscle fiber, a bacterial leak from the gut—into information that other cells can read and act on. The redness and swelling are not the message itself; they are the visible side effects of the message being broadcast. Inflammation is a data transmission protocol implemented in chemistry: molecules as bits, gradients as wires, time as encoding.

A useful starting point is to notice that tissues are normally quiet. White blood cells patrol, but most of the time they are not activating, secreting, dividing, or killing. The body cannot afford to run in emergency mode as a baseline. When something goes wrong, a tissue must answer a hard question quickly: does this require a silent cleanup, or should it recruit an entire army? That decision is not made by a single cell or molecule. It emerges from a conversation among many players, each with partial information, each using a shared molecular vocabulary.

The vocabulary is mostly made of cytokines and chemokines. Cytokines are small protein signals released by cells that change the behavior of other cells. Chemokines are a subset specialized for directing cell movement, forming trails that immune cells can follow like scent. The same cell can release multiple cytokines at once, and the meaning of any one cytokine depends on context: what else is present, how much, for how long, and which cells are listening.

This is why the phrase "cytokine storm" became so widely used in public conversation in the early 2020s. It evokes chaos, a system overwhelmed by its own signals. But the metaphor also points to a deeper truth: even in ordinary inflammation, the immune system relies on interlocking, amplifying messages that can become dangerous if they fail to shut off. A storm is a failure of control, not a different kind of weather.

The mechanics begin at the site of trouble, with the hardwired sensors described earlier. A macrophage in the lung does not merely *detect* a threat; it translates detection into a pattern of secreted signals. A dendritic cell does the same, often adding a crucial twist: it can carry a molecular snapshot of what it has encountered to a lymph node, where adaptive immunity can be recruited. Epithelial cells lining the airway or gut, once thought of as passive barriers, also act as signal broadcasters. They can release cytokines that shape inflammation long before professional immune cells arrive in force.

The first wave of signals is often local and practical. Blood vessels widen (vasodilation), increasing blood flow and causing redness and warmth. Vessel walls become leakier, letting fluid and proteins enter tissue, causing swelling. These changes look like disorder, but they are delivery mechanisms. More blood means more immune cells can arrive. Leaky vessels mean antibodies, complement proteins, and clotting factors can access the site. Swelling increases the tissue's capacity to host the response, like opening lanes on a highway during rush hour.

Pain is also part of the protocol. Nerve endings respond to inflammatory mediators; pain discourages movement and encourages protection of the affected area. This is not comforting to the person experiencing it, but it can be adaptive. Pain is a behavior-modifying signal produced by immune chemistry.

The most interesting part, though, is what cannot be seen: the message content embedded in the particular mix of cytokines. A macrophage encountering bacterial fragments may release TNF (tumor necrosis factor) and IL-1, both strong drivers of inflammation, along with IL-6, which promotes fever and the production of acute-phase proteins by the liver. It may also release chemokines that summon neutrophils. A virus-infected cell, in contrast, may emphasize type I interferons, which warn neighboring cells to raise antiviral defenses and help coordinate cytotoxic responses. These are not fixed rules; they are tendencies shaped by receptor engagement, cell type, tissue location, and previous history.

This is where inflammation becomes a decision layer rather than a mere byproduct. Cytokines do not just "turn immunity on." They steer it. They determine which cells dominate a response, which weapons are drawn, how much collateral damage is tolerated, and how quickly the system attempts to return to baseline.

To see inflammation as information, consider a simple thought experiment. Imagine two identical cuts on a finger. In one, only sterile saline enters the wound. In the other, a small number of bacteria enter. Both injuries involve tissue damage, so both will produce some local inflammation: damaged cells release so-called danger signals, and clotting begins. But the bacterial contamination adds specific molecular cues that shift the signal mixture. The tissue is no longer reporting merely "damage" but "damage with microbial patterns." That distinction changes the recruitment profile, the intensity of the response, and the likelihood that adaptive immunity will be enlisted. The difference is not philosophical. It is encoded in molecules, read by receptors, and amplified by cytokines.

Communication systems need both amplifiers and brakes. The immune system uses feedback loops for both. TNF and IL-1 can induce cells to produce more cytokines, which in turn recruit and activate more immune cells that produce still more cytokines. This positive feedback can be lifesaving when a rapidly dividing bacterium threatens to spread. The same logic can also produce runaway inflammation if the trigger is not contained, if the brakes fail, or if the system mistakes the trigger.

The brakes are not a single off-switch. They include anti-inflammatory cytokines such as IL-10, regulatory cell types, receptor desensitization, and physical removal of stimulus through clearance of microbes and dead cells. There are also more subtle forms of braking: some cytokines promote responses that compete with other responses, and the body's stress hormones can dampen inflammation systemically. The overall effect is that inflammation is constantly negotiating between two risks: underreaction, which allows infection or damage to spread; and overreaction, which harms tissue and can endanger the whole organism.

A historical anecdote helps anchor this abstract picture. In the nineteenth century, inflammation was a battleground for ideas about disease. Rudolf Virchow, one of the founders of modern pathology, emphasized that disease processes could often be understood at the level of cells and tissues. The classical signs of inflammation were part of medical education long before cytokines were known. What changed in the late twentieth century was the discovery that the signs were driven by a language of soluble factors. The very word "cytokine" is relatively modern, and the mapping of specific molecules to specific effects became a major scientific enterprise. Clinicians learned that fever is not simply heat leaking from an infection but a regulated change in the hypothalamic thermostat, driven by cytokines such as IL-1 and IL-6 that induce prostaglandins. The body was not passively suffering fever; it was choosing it, at a cost, because higher temperature can hinder some pathogens and enhance certain immune functions.

Fever illustrates the protocol idea especially well because it makes the local systemic. A small infection in a lung lobe can lead to whole-body temperature change. That is a broadcast decision: the immune system is willing to spend energy and cause discomfort because it judges that a systemic adjustment will help. When the decision is wrong or excessive, fever can become dangerous. But the existence of fever as a controlled response tells us that inflammation is not merely local chaos.

Some of the most dramatic examples of inflammatory miscalibration occur in cytokine storm syndromes, a term used to describe hyperinflammatory states in which high levels of cytokines lead to systemic symptoms and organ dysfunction. The causes vary: severe infections, certain autoimmune conditions, and, notably, some immunotherapies designed to unleash T cells against cancer. In cancer therapy, chimeric antigen receptor (CAR) T cells can be spectacularly effective, and they can also trigger cytokine release syndrome, a storm driven not by a microbe but by massive immune activation. The patient may develop high fever, low blood pressure, and breathing difficulty. Treating the storm sometimes involves blocking particular cytokines, such as IL-6 signaling, while still trying to preserve the anti-cancer effect.

This clinical phenomenon underlines two important points. First, inflammation is powerful enough to harm the host even when no pathogen is directly causing the damage. Second, a storm is not a random explosion; it often has recognizable features and leverage points. Blocking IL-6 or IL-1 can sometimes calm the system. These targeted interventions would be impossible if inflammation were merely tissue damage. They are possible because inflammation is structured signaling.

That structure is clearer when one looks at how inflammatory signals are produced inside cells. A PRR on the surface or in an endosome activates intracellular signaling pathways that converge on transcription factors. These transcription factors enter the nucleus and switch on genes for cytokines, chemokines, and other immune effectors. The cell's response is therefore partly a gene-expression program, not just an instantaneous reflex. Different receptors share some signaling modules, which is efficient, but they also differ in adaptor proteins and pathway branches, which allows different inputs to produce different outputs. Cells are not computing with silicon, but they are integrating signals through biochemical networks that have logic-like properties: thresholds, gating, amplification, and timing.

Timing deserves special emphasis. A single cytokine pulse can mean something different from sustained cytokine production. Early TNF may help recruit immune cells and activate endothelium (the lining of blood vessels). Prolonged TNF can contribute to chronic inflammation and tissue breakdown. The immune system uses time as an axis of meaning, in the same way that a short beep and a continuous alarm carry different urgency.

A second axis is location. Cytokines act locally at low concentrations and systemically at high concentrations. A small amount of IL-1 in tissue may recruit leukocytes without altering body temperature. Enough IL-1 and IL-6 entering the bloodstream can induce fever and acute-phase responses from the liver. This scaling property is one reason inflammation can be both appropriate and dangerous: a response that is helpful when local can become harmful when it spills into systemic circulation.

A third axis is cellular audience. The same cytokine can have different effects on different cell types because those cells express different receptors and different downstream machinery. Interferons can induce an antiviral state in epithelial cells, but they can also enhance antigen presentation in dendritic cells and influence the differentiation of T cells. IL-10 can calm macrophages and reduce inflammatory cytokine production, but its effects depend on the cell's activation state and environment. This is not a flaw; it is a way to reuse molecular words in different grammatical contexts.

Inflammation also contains a fascinating element of quality control: it is not simply "on" or "off." Some inflammatory pathways require multiple signals, effectively asking for corroboration before releasing particularly dangerous messages. The NLRP3 inflammasome is a well-studied example. It is a molecular machine inside cells that, when assembled, activates an enzyme (caspase-1) that processes IL-1β into its active form and can trigger a fiery form of cell death. IL-1β is a potent cytokine: extremely useful in host defense, extremely capable of causing harm.

The common description of NLRP3 activation involves two signals. The first is priming: a warning that prepares the cell by increasing the production of NLRP3 and pro-IL-1β, often through NF-κB activation by PRRs or cytokines. The second is activation: evidence of cellular disturbance, such as changes in ion concentrations, extracellular ATP signaling tissue damage, or particulates that disrupt lysosomes. Only when the cell has both preparedness and disturbance does it commit to assembling the inflammasome and releasing mature IL-1β. In communication terms, this is a safeguard against false positives. It is harder to trigger by accident than a single-switch system.

This two-signal logic also shows how the immune system folds physical reality into chemical meaning. ATP is the cell's energy currency, normally kept inside cells. ATP outside cells can signal cell injury. Crystals or particulates that damage lysosomes indicate a kind of insult that can accompany infection or sterile injury. The inflammasome is sensitive to these perturbations, treating them as evidence that the situation is not merely a passing molecular whiff but a meaningful disruption requiring escalation.

Of course, escalation must still be controlled. IL-1 pathways are implicated in inflammatory diseases, and medications that block IL-1 signaling can relieve certain conditions dramatically. This is a reminder that the same molecules that protect against infection can become pathogenic when amplified or prolonged. The immune system cannot change its basic alphabet every time it encounters a new situation; it must reuse powerful signals in a wide range of contexts and rely on regulation to prevent harm.

Everyday experience contains small demonstrations of this reuse. A sore throat from a virus and a swollen ankle after a sprain can both involve similar inflammatory mediators. One is caused by infection, the other by mechanical injury. The immune system reads both as situations where barriers have been breached and tissue integrity is threatened, and it responds with overlapping modules: increased blood flow, recruitment of immune cells, pain, and a cleanup program to remove debris and promote repair. The overlap can be frustrating—the ankle hurts even though no microbe is present—but it reflects a deep evolutionary economy. A single set of mechanisms handles multiple classes of problems by responding to patterns of damage and danger.

Inflammation also shapes perception and behavior beyond pain. Systemic inflammatory signals can cause fatigue, reduced appetite, and changes in sleep. These "sickness behaviors" are not merely consequences of being ill; they are partly regulated responses that may conserve energy for immune work and reduce exposure to new threats. The immune system, through cytokines, can influence the brain's priorities. Again, this is not a mystical mind within the immune system. It is communication across tissues using molecules that have receptors in the nervous system and influence neural circuits.

One might ask: if inflammation is a protocol, who designed it? Evolution did, but not as an engineer would. It is a patchwork assembled over hundreds of millions of years, with redundancies and quirks. Some cytokines do similar jobs because they arose at different times or in response to different pressures. Some pathways are surprisingly indirect because they are built on older circuitry. Yet the overall effect is coherent enough to save lives daily.

A final way to grasp the sophistication is to think about what inflammation must accomplish simultaneously. It must fight microbes, limit their spread, repair tissue, and avoid provoking self-destruction. These goals conflict. Killing infected cells can damage organs. Recruiting neutrophils can clear bacteria but also release enzymes that harm surrounding tissue. Fever can hinder pathogens but strains the heart and dehydrates the body. Inflammation is the process by which these trade-offs are negotiated in real time.

When the negotiation works, the tissue returns to quiet. Blood vessels tighten back to normal, excess fluid drains through lymphatics, dying neutrophils are cleared by macrophages, and repair programs rebuild what was damaged. The sore throat resolves, the ankle swelling subsides, the fever breaks. The visible drama fades because the message has done its work and the broadcasting stops.

When it fails, the same communication system that once coordinated protection can become a loop that keeps feeding itself: cytokines inducing more cytokines, recruiting cells that amplify the message, sustaining pain and swelling long after the original trigger is gone. Chronic inflammation is, in a sense, a protocol that has lost its stop condition.

Inflammation is therefore neither friend nor enemy. It is a language of urgency, a set of signals that can be tuned, misread, overheard, or amplified. To feel inflammation as mere discomfort is understandable; it is the body demanding attention. But beneath the discomfort is an intricate act of coordination: cells telling other cells where trouble is, what kind it might be, and how much force to bring to bear, all while trying to leave the tissue livable when the emergency is over.

2.3 The Epigenetic Ghost

A person who has recovered from chickenpox usually does not get it again. That sort of lasting protection is the familiar triumph of immune memory, written into the long-lived populations of B cells and T cells that remember a specific pathogen with unnerving clarity. For a long time, this success story cast a long shadow. If memory meant specificity, and specificity meant antigen receptors made by gene rearrangement, then memory belonged to the adaptive immune system by definition. The innate immune system, in that older telling, could be fast and forceful, but it could not learn.

Reality has turned out to be more interesting, and more unsettling. Some of the body's most ancient defenders, cells that carry hardwired sensors rather than bespoke receptors, can change their future behavior based on past encounters. They do not remember a particular viral epitope the way a memory B cell does. Their memory is broader, blurrier, and sometimes paradoxical: after one challenge, an innate cell may respond more fiercely to a later and unrelated challenge; after another, it may respond less, as if chastened. These are not vague impressions. They are measurable shifts in cytokine production, microbial killing, and gene expression.

The term *trained immunity* has become the shorthand for one of these shifts: a durable increase in innate responsiveness after an initial stimulus. The word *trained* is carefully chosen. It avoids claiming that monocytes, macrophages, or natural killer cells possess the kind of fine-grained, antigen-specific memory that T and B cells do, while still insisting that the innate system is not merely a reflex. Training implies a changed baseline: a different readiness to respond, a different threshold for alarm, a different capacity to produce the molecules that coordinate inflammation.

This is a provocative idea because many innate immune cells are short-lived. Neutrophils survive for days. Circulating monocytes last longer but still turn over on the scale of weeks. How can such cells "remember" anything for months? The answer, as so often in biology, is that the unit of memory is not necessarily the individual cell. Sometimes it is the lineage: a change imprinted in precursor cells in the bone marrow, or a lasting alteration in tissue-resident macrophages that are replenished slowly. And sometimes, even within a cell that does not live long, a molecular change can persist long enough to influence a later encounter.

The mechanism that makes this plausible is neither mystical nor metaphoric. It lives in the cell nucleus, in the physical state of DNA packaging. Human DNA is not stored as a loose thread. It is wrapped around proteins called histones, forming a structure known as chromatin. Chromatin can be tight and relatively inaccessible, like a book sealed shut, or it can be open, like pages fanned and ready. A gene that is wrapped tightly is hard to transcribe. A gene that sits in open chromatin is easier to turn on quickly.

Cells control this accessibility with chemical tags placed on histones and sometimes on DNA itself. These tags do not change the genetic code; they change how the code is read. That is the essence of *epigenetics*: regulation "above" the genome, governing expression rather than sequence. Epigenetic regulation is not unique to immunity. It shapes development, cell identity, and the stable differences between a neuron and a liver cell even though they contain the same DNA. What is striking is that innate immune cells can use the same machinery to store a trace of past inflammation.

An infection can push a macrophage into an activated state: genes for inflammatory cytokines, antimicrobial enzymes, and chemokines are switched on. When the infection resolves, the cell does not necessarily return to exactly the same chromatin landscape it had before. Some regions remain more open than they were. Certain enhancers—DNA elements that act like volume knobs for gene expression—can stay "primed," marked by histone modifications that make them easier to reactivate. The cell looks calm, but a molecular ghost of the encounter lingers in its chromatin, shaping its future reactions.

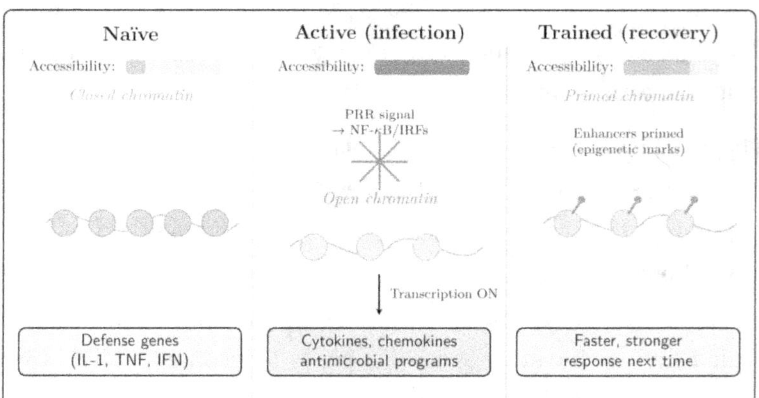

Color intensity indicates chromatin accessibility: darker/tighter means harder to read; lighter/open means ready.

The diagram is deliberately abstract because the core idea is architectural: an innate cell can be left, after an encounter, with a genome that is easier to read in certain places. A second encounter then produces a faster or higher-amplitude response, not because the cell has acquired a new receptor, but because the transcriptional machinery can reach key genes more quickly.

Trained immunity did not emerge from a single eureka moment so much as from accumulating puzzles. One puzzle was epidemiological and lived in the messy world of human populations: some vaccines seemed to have benefits beyond their target pathogen. The Bacillus Calmette–Guérin (BCG) vaccine, developed in the early twentieth century to protect against tuberculosis, has long been notable for its broad, sometimes controversial associations with reduced mortality from infections not caused by *Mycobacterium tuberculosis*. Many factors can confound such observations, and immunologists are rightly cautious about turning population patterns into mechanistic claims. Still, the idea that a vaccine could "tune" innate defenses helped motivate experiments that asked a simpler question: after BCG vaccination, do innate immune cells respond differently to unrelated stimuli?

Work associated with Mihai Netea and colleagues in the 2010s helped bring this into focus. In human studies, monocytes collected months after BCG vaccination could produce more cytokines when stimulated ex vivo with microbes unrelated to tuberculosis. Importantly, this enhanced responsiveness did not depend on generating new antigen-specific receptors in those monocytes; it correlated with epigenetic changes at inflammatory gene loci, including increased levels of histone marks such as H3K4me3 that are often associated with active or activatable chromatin. The immune system, it seemed, could keep a molecular memory in places other than the rearranged receptors of lymphocytes.

A second line of evidence came from a component of fungal cell walls called β-glucan. Expose monocytes to β-glucan, let them rest, then stimulate them again later with a different trigger, and the second response can be stronger than it would have been without the first exposure. This is training in a controlled dish: an initial encounter that leaves a lasting bias.

The next question is inevitable. Epigenetic changes sound like scribbling in the margins of the genome, but scribbles still need ink. Where does the cell get the energy and raw materials to reconfigure chromatin and sustain a heightened state of readiness?

Here metabolism enters, not as a background housekeeping chore but as a partner in immune decision-making. When an innate immune cell becomes activated, it often shifts its metabolic strategy. It may rely more on glycolysis—the rapid breakdown of glucose—even when oxygen is available. This is reminiscent of the "Warburg effect" described in cancer cells, and it reflects a trade-off: glycolysis is less efficient per glucose molecule than oxidative phosphorylation, but it can provide energy and biosynthetic intermediates quickly. Rapid growth and rapid cytokine production demand building blocks.

Training appears to depend on such metabolic reprogramming. Pathways involving mTOR, Akt, and HIF-1α have been implicated in β-glucan-induced training, linking nutrient sensing to the immune state. Metabolites are not just fuel; they can influence chromatin-modifying enzymes. Some histone modifications require specific metabolites as substrates or cofactors. In that sense, metabolism provides both the energy and the chemical currency for writing epigenetic marks.

The language of training, however, can mislead if it suggests that innate responses simply get stronger with experience in a uniformly beneficial way. The innate system also learns the opposite lesson.

After a strong inflammatory exposure, particularly to bacterial LPS, innate cells can become less responsive on subsequent stimulation. This is often called *endotoxin tolerance*. In clinical settings, a tolerance-like state is sometimes invoked to explain aspects of immune paralysis seen after severe sepsis, where patients become vulnerable to secondary infections. The term "tolerance" can be confusing because immunology uses it in several ways, including the adaptive immune system's tolerance to self. Here the meaning is narrower: a dampened innate inflammatory response after a prior inflammatory stimulus.

Tolerance is not merely exhaustion. It can be selective. Some genes become harder to activate, while others remain inducible. A cell might reduce production of certain pro-inflammatory cytokines while maintaining antimicrobial functions. This selectivity again points to chromatin. Rather than flipping a global off-switch, the cell appears capable of remodeling accessibility gene by gene, silencing some programs while keeping others available.

This makes intuitive sense if inflammation is treated as a protocol. A system that cannot recalibrate after a false alarm is prone to self-harm; a system that cannot escalate after a genuine attack is prone to infection. Training and tolerance are two modes of recalibration. Both can be adaptive in the right context and harmful in the wrong one.

Consider the lived experience of the immune system in a modern city. People encounter microbial products constantly: on food, in dust, on skin, in the gut. If every exposure triggered maximum inflammation, life would be impossible. The immune system must ignore most signals or respond quietly. Yet when a barrier is breached and microbes enter tissue, it must respond quickly and decisively. The ability to tune future responsiveness based on past encounters is therefore not a luxury; it is a way of surviving in a world where danger is frequent but not constant.

This tuning also has an uncomfortable implication: experiences outside of infection can plausibly shape innate immune memory. Chronic stress, diet, pollutants, and tissue damage might all influence inflammatory set points through metabolic and epigenetic routes. The evidence in humans is complex and often correlational, but the mechanistic possibility is clear. If chromatin marks can prime inflammatory genes, then environments that chronically push the system toward inflammation might leave lasting marks as well.

The cellular level offers one more twist that makes the "ghost" metaphor apt. Many trained-immunity effects outlast the cells first exposed to the training stimulus. One explanation is that training can occur not only in mature circulating monocytes but also in hematopoietic stem and progenitor cells in the bone marrow, biasing the next generations of myeloid cells toward a heightened state. In that scenario, the memory is not carried by a single long-lived cell but by a changed production line. New cells are born already inclined to respond in a particular way.

Another explanation is that some innate cells are longer-lived than their reputation suggests. Tissue-resident macrophages, for instance, can persist for long periods and self-renew locally. They are shaped by their tissue environment and, once trained or tolerized, may continue to imprint their biases on local inflammation.

It is tempting to treat trained immunity as an attractive biological hack: a way to boost defense broadly without needing pathogen-specific vaccines. There is genuine promise here. Training might enhance resistance to certain infections, or improve responses in people with weakened immunity. But the dangers are real, too. A system with a lower threshold for inflammation may be more likely to damage tissues, contribute to atherosclerosis, worsen autoimmune diseases, or amplify cytokine storms when a severe infection hits. The immune system's power is never free.

A concrete case shows the knife-edge. Imagine two people exposed to the same respiratory virus. One has an innate immune system that produces an early, appropriately scaled interferon response in the airway, limiting viral replication. The other produces a delayed interferon response but a large late inflammatory burst. These patterns have been discussed in various viral infections, including severe respiratory disease, as possible contributors to different outcomes. The point is not that "more inflammation" is always worse; it is that timing and balance matter. A trained state that accelerates useful early programs could be protective, while a trained state that amplifies late systemic inflammation could be dangerous. Epigenetic priming does not come with a moral compass. It makes certain responses easier to trigger; whether that is good depends on circumstance.

The phrase "epigenetic scars" is sometimes used to describe these chromatin changes. Scars are a helpful image because they capture both persistence and compromise. A scar marks a past injury and can strengthen tissue, but it can also reduce flexibility. Chromatin scars can similarly make a gene region more responsive, or they can lock it into a less responsive state. Either way, the genome carries a trace of experience, not in its sequence but in its accessibility.

One of the most intriguing findings in this area is that these states can sometimes be reversed or rewritten. In experimental settings, cells rendered tolerant by one stimulus can have responsiveness restored by another, accompanied by changes in chromatin marks at key regulatory elements. That reversibility is both hopeful and sobering. It suggests potential therapies that could retune harmful inflammatory memories, but it also suggests that the immune system's set points are not fixed even in adulthood; they are negotiable, and therefore vulnerable to the pressures of environment and disease.

What remains constant through all of this is the fundamental difference between adaptive and innate memory. Adaptive memory is like keeping a mugshot: a precise record of a particular pathogen's features, enabling targeted responses upon re-encounter. Trained immunity is more like adjusting the sensitivity of the alarm system after a break-in. The cameras do not learn the burglar's face, but the motion sensors become quicker to trigger, and the dispatch system responds with greater urgency. Sometimes that is exactly what is needed. Sometimes it leads to a neighbor calling the police because a cat walked past a window.

The innate immune system, then, is not a static wall and not merely a set of reflexes. It is a living interface that can be tuned by experience, with chromatin as its memory substrate and metabolism as both engine and ink. When a macrophage or monocyte seems to "remember," it is not reminiscing. It is reading a genome whose pages have been subtly rearranged, margins annotated, and chapters left open at places that once mattered. The next time the cell detects trouble, the response can arrive not only fast, but already leaning in a particular direction, as if the past has not quite finished speaking.

Chapter 3

The Architects of Specificity

Biological defense requires more than brute force; it demands the ability to distinguish a harmless stranger from a lethal pathogen with molecular exactitude. This transition from general surveillance to targeted warfare relies on a sophisticated relay of intelligence, where specialized couriers physically transport evidence of invasion to the body's command centers to unlock the arsenal of adaptive immunity.

3.1 The Intelligence Handoff

A cut on a finger is not, by itself, a complicated problem. Bacteria land where they should not be, the tissue protests, blood clots, and within hours the mess is mostly contained. Yet the immune system behaves as though every cut might be a rehearsal for something far worse: a virus quietly turning cells into copy machines, a parasite hiding behind the body's own molecules, a tumor learning how to look ordinary. The early stages of defense are fast and blunt, but the immune system also needs something slower and exquisitely discriminating: cells that can recognize a particular molecular signature and remember it for years.

The handoff between these two tempos—the immediate and the

precise—is one of immunology's great pieces of choreography. It is performed by dendritic cells, a family of sentinels that do more than detect trouble. They carry evidence from the scene, decide whether it matters, and present it to the small, quiescent lymphocytes that are capable of building a tailored response. They do not simply deliver information; they translate it into a decision.

A useful way to picture dendritic cells is as professionals of context. Many immune cells can swallow debris. Many can secrete inflammatory chemicals. Many can kill. Dendritic cells specialize in something subtler: they take fragments of the outside world and mount them like exhibits in a courtroom, while simultaneously reporting the circumstances under which those fragments were found. Was there tissue damage? Were there viral alarm signals inside the cell? Was bacterial cell wall detected in the neighborhood? Were stress molecules released by dying cells? In immunological terms, dendritic cells connect *antigen* (what is there) with *danger* (what it means).

The significance of this became clear only relatively recently. In 1973, Ralph Steinman and Zanvil Cohn described a strange, branching cell in the spleen that did not fit comfortably into existing categories. The cell's delicate projections suggested a tree in winter, and they named it for that shape. For years, dendritic cells were an oddity: rare, difficult to grow, easy to overlook. The turning point was the realization that these cells were extraordinarily good at activating naive T cells—the inexperienced T cells that have not yet been committed to a particular enemy. That single fact reorganized how immunologists understood the start of adaptive immunity. It also carries a human lesson: a discovery can be transformative not because it adds another piece, but because it shows what a piece *is for*.

To appreciate what dendritic cells do, it helps to start with what they look like when they are *not* doing it. In many tissues, dendritic cells spend their days in a kind of sampling mode. They extend processes between epithelial cells, patrol the spaces under the skin and along mucosal surfaces, and constantly ingest bits of their surroundings. This does not necessarily mean the tissue is inflamed. A healthy intestine, for example, is full of foreign material—food proteins, trillions of bacteria—and an immune system that reacted strongly to every molecule would be unlivable. Much of the time, dendritic cells in such settings behave as guardians of restraint. They can present antigens in ways that teach T cells to ignore harmless targets or to adopt regulatory roles that dampen inflammation. The same machinery that can ignite immunity can also enforce tolerance.

What flips a dendritic cell from sampler to messenger is not simply *finding* antigen. It is detecting the signatures that suggest an infection or injury. These signatures come in two broad types. One consists of molecular patterns associated with microbes: components of bacterial cell walls, viral RNA in the wrong compartment, unusual DNA motifs, fungal carbohydrates. The other consists of the body's own distress signals: molecules released by damaged cells, ATP spilling into the extracellular space, changes in ionic balance, fragments of extracellular matrix that indicate tearing. The receptors that sense these cues—a collection that includes Toll-like receptors and other pattern-recognition systems—are often described as belonging to innate immunity. That label is accurate, but incomplete. Their deeper function is to give dendritic cells a way to interpret a scene.

Once those receptors are engaged, a dendritic cell begins to change its priorities. The shift is sometimes called *maturation*, but that word can mislead; it is not about becoming older or more developed, but about becoming oriented toward communication. A maturing dendritic cell tightens up how it processes and displays antigen. It increases the expression of molecules needed for T-cell activation, particularly co-stimulatory proteins such as CD80 and CD86. It alters the set of chemokine receptors it carries, most notably upregulating CCR7, which makes the cell responsive to chemokine trails leading to the T-cell zones of lymph nodes. It also changes its secretions: different dendritic cell states release different cytokines, and these cytokines will later influence what kind of T-cell response is built.

Then comes a physical journey that is easy to underestimate. The infection might be in the skin of the forearm, the lining of the lung, or the tissue around the gut. The T cells that need to be recruited are not waiting there in large numbers. They are mostly circulating, passing through lymph nodes like travelers through train stations. If the immune system relied on chance encounters in peripheral tissues, it would be slow and unreliable. Instead, dendritic cells carry the problem to a place designed for meetings.

The lymphatic system is that meeting infrastructure. Fluid from tissues drains into lymphatic vessels, carrying soluble molecules and sometimes whole microbes. Dendritic cells can enter these vessels and travel to the nearest draining lymph node. The route is not a simple slide down a pipe; it is guided by molecular gradients and constrained by anatomy. Within the lymph node, dendritic cells and T cells follow distinct paths into the interior, and their movement is organized so that a rare T cell with a matching receptor has a reasonable chance of finding its cognate dendritic cell. This is one of the immune system's recurring strategies: solve a search problem not by making each searcher infinitely smart, but by arranging the environment to make successful encounters more likely.

A lymph node is not merely a bag of cells. It is compartmentalized. There are areas rich in B cells, which are set up to sample antigen in a different way, and there is a paracortical region rich in T cells. Dendritic cells that arrive from tissues are drawn into the T-cell zone, where they spread their branching processes and display peptide—

MHC complexes like many small billboards. Naive T cells enter the node, migrate through that same zone, and repeatedly make brief contacts with dendritic cells. Most of these contacts are unproductive. A T cell's receptor is exquisitely specific, and any single T cell can recognize only a tiny subset of possible peptides. The immune system compensates by sheer diversity: many different T cells, each with a distinct receptor, continuously scanning.

At the heart of the handoff is a molecular device that is both elegant and strange: the major histocompatibility complex, or MHC. A dendritic cell does not present whole proteins to T cells. Instead, it breaks proteins into short peptides and loads those peptides onto MHC molecules. The MHC then carries the peptide to the cell surface. The peptide is the evidence; the MHC is the display case.

There are two main classes. MHC class I presents peptides generally derived from proteins in the cell's cytosol—often a way of displaying what is happening *inside* a cell, including viral infection. MHC class II presents peptides derived from proteins taken up from outside the cell and processed in endosomal compartments—a way of displaying what is happening in the *environment*. The split matters because it aligns with two broad T-cell populations. CD8 T cells tend to recognize peptides on MHC I; CD4 T cells tend to recognize peptides on MHC II. But dendritic cells have an additional trick, vital for antiviral defense: they can sometimes take material from outside and present it on MHC I, a process called *cross-presentation*. This allows the immune system to activate CD8 T cells even when the dendritic cell itself is not infected.

The encounter between a dendritic cell and a T cell is often described as an *immunological synapse*, borrowing language from neuroscience. The analogy is not perfect, but it captures two key ideas: physical intimacy and organized signaling. This is not a diffuse chemical conversation across a room. The two cells press membranes together and build a structured interface in which receptors, adhesion molecules, and signaling complexes cluster in patterns. The synapse stabilizes contact long enough for a T cell to decide whether what it is seeing is meaningful.

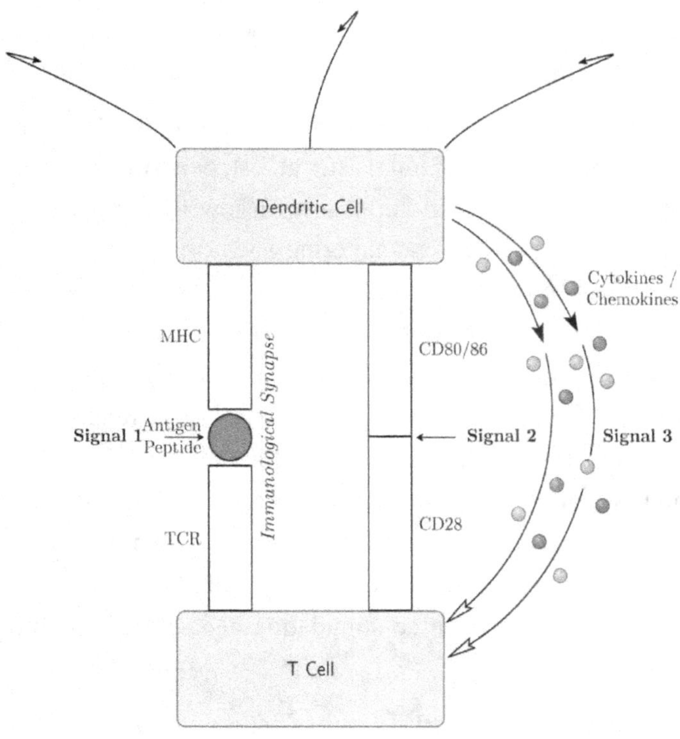

The synapse helps clarify a crucial point: recognition alone is not enough. A T cell that receives only the first signal—its receptor binding a peptide—MHC complex—is in danger of being activated by the wrong thing. The body is full of self-derived peptides, and a certain amount of accidental recognition is unavoidable. Dendritic cells reduce this risk by making full activation conditional on additional information.

Immunologists often describe three categories of input for a naive T cell. The first is antigen-specific recognition: the T-cell receptor binds a particular peptide presented by a particular MHC. The second is co-stimulation: proteins on the dendritic cell surface engage receptors such as CD28 on the T cell, providing a permissive signal that says, in effect, *this is worth responding to*. The third is the cytokine environment: soluble factors that influence what type of effector the T cell becomes, shaping the character of the response.

Co-stimulation is where the dendritic cell's translation role becomes most visible. In peripheral tissues at rest, dendritic cells may present self and environmental antigens with low levels of co-stimulatory molecules. A T cell encountering such presentation may become unresponsive or may differentiate into a regulatory phenotype that discourages inflammation. When danger receptors are engaged, dendritic cells upregulate CD80 and CD86 and other activating ligands, tipping the balance toward immunity. This arrangement is sometimes called the "two-signal" requirement, and it is one of the immune system's fundamental safety interlocks. It is also a place where medicine can intervene. Drugs that block co-stimulation can dampen unwanted immune responses in transplantation or autoimmunity; therapies that enhance co-stimulation can strengthen antitumor responses.

Cytokines provide a third layer of interpretation. A dendritic cell exposed to intracellular viral signals may produce type I interferons and other cytokines that favor strong cytotoxic T-cell responses. In settings that resemble helminth infection, other signals dominate, and T cells differentiate into helper types more suited to mobilizing eosinophils and mast cells. The dendritic cell is not simply delivering antigen; it is delivering a verdict about the kind of world that antigen comes from.

A concrete example makes the translation vivid: consider vaccination, particularly those that include adjuvants. An adjuvant is often described as an "immune booster," but that phrase hides the logic. Many purified antigens are poor at provoking immunity because they are, immunologically speaking, too polite. A protein fragment alone may be seen as information without urgency. Adjuvants provide the missing context by engaging innate sensors and provoking dendritic cell maturation. This is not merely turning up the volume. It is changing the message from *here is a molecule* to *here is a molecule associated with danger, and here is the kind of response that fits*. The success of many vaccines depends less on the antigen being visible and more on dendritic cells being convinced that it matters.

The same principle is exploited in cancer immunotherapy, where one of the central problems is that tumors often look like self. Even when tumors carry mutations that generate novel peptides, those peptides may be presented in contexts that lack strong co-stimulation. Some modern therapies aim to reverse that. Checkpoint blockade, for example, does not directly activate dendritic cells, but it removes inhibitory brakes on T cells, making it easier for activation signals to succeed. Other approaches attempt to deliver tumor antigens to dendritic cells in an inflammatory context or to stimulate dendritic cells directly within tumors. The details vary, but the underlying need is consistent: antigen without an appropriate handoff is often biologically ignored.

There is a deeper, slightly unsettling implication to all of this. The immune system's celebrated specificity does not arise simply because T cells and B cells have exquisitely shaped receptors. It arises because specificity is *licensed*. A T cell may have a receptor that fits a peptide like a key fits a lock, yet still be instructed not to act if the lock appears in a peaceful neighborhood. Dendritic cells are the officials who issue that license.

This is also why some infections are dangerous in ways that go beyond their immediate damage. A pathogen that can disrupt dendritic cell function can derail the entire adaptive response. Some viruses infect dendritic cells or interfere with their maturation, blunting co-stimulation and cytokine production. Others manipulate cell death pathways so that antigen is presented in tolerogenic forms. The immune system's dependence on translation creates vulnerabilities, just as any system that relies on trusted messengers does.

The handoff ends, in a sense, when a T cell commits. After sufficient productive synapses, a naive T cell begins to proliferate, generating a clone of cells with the same receptor. The commitment is visible under the microscope as a once-small, quiet lymphocyte becomes a larger, metabolically active cell, dividing again and again. The lymph node, which was built to maximize rare encounters, becomes a factory for expansion. Yet the decision that triggers this expansion is not made by the T cell alone. It is made in partnership with dendritic cells that have brought antigen, interpreted danger, and presented both in a structured conversation at the synapse.

If immune memory is the body's way of learning from experience, dendritic cells are among its most influential teachers. They choose which lessons count, and they teach by showing not only *what* was found, but *how* it was found: in calm, in chaos, in the presence of damage, or amid the subtle chemical signatures of an intracellular invasion. A dendritic cell arriving in a lymph node carries, quite literally, a piece of the world—and, more importantly, the meaning it assigns to that piece.

3.2 Commanders and Executioners

The first time a naive T cell recognizes its match on a dendritic cell, it is not yet a soldier of any particular kind. It is more like a recruit handed a file folder and asked to decide whether the case is real, urgent, and within its jurisdiction. The decision is physical: receptors bind, membranes press close, internal scaffolding rearranges, and a quiet cell begins the expensive business of growth. Then comes a second decision that matters just as much as recognition itself: what sort of T cell to become.

Two broad fates dominate this story. Some T cells mature into CD4 *helper* T cells, coordinators that manage other immune cells through contact and chemical instruction. Others become CD8 *cytotoxic* T cells, executioners that can directly kill infected or damaged cells. The division is so fundamental that it shows up in the architecture of the immune system: helper T cells are trained to read peptides displayed on MHC class II molecules, while cytotoxic T cells read peptides on MHC class I. The immune system, in other words, separates the world into two kinds of problems: threats that are largely *outside* cells (and so must be handled by organized, multi-cell cooperation) and threats that are *inside* cells (and so require the surgical removal of compromised cellular factories).

This is a simplification, but it captures the logic. Bacteria that live in tissues, parasites that move through blood, and toxins released into the extracellular space call for antibodies, complement, and phagocytes—tools that work best when directed and amplified by helper T cells. Viruses, by contrast, turn ordinary body cells into production sites. Once a virus has entered a cell, antibodies can no longer reach it. The immune system needs a way to identify infected cells and stop them, even if they look outwardly normal. That is the domain of CD8 cytotoxic T cells.

The distinction between CD4 and CD8 is visible on the surface of the cells. CD4 and CD8 are co-receptors: they sit near the T-cell receptor (TCR) and help it signal effectively. CD4 binds to MHC class II; CD8 binds to MHC class I. Yet the deeper distinction is not the labels, but the specialization of labor. Helper T cells are influential because they can *license* other cells to act. Cytotoxic T cells are dangerous because they can kill.

This bifurcation begins at the very moment of contact with a dendritic cell. The immunological synapse is not just a handshake; it is an active construction project. Signaling proteins cluster into microdomains, kinase cascades are triggered, and transcriptional programs wake up. If a T cell is a decision-making machine, the synapse is its control panel.

A few molecular names illustrate what is happening without requiring one to memorize them. When the TCR engages a peptide—MHC complex, it recruits and activates a kinase called Lck, which then helps activate another kinase, ZAP-70. These in turn phosphorylate adaptor proteins, notably LAT and SLP-76, which function like docking stations: once modified, they assemble larger signaling complexes that amplify and shape the message. The result is not a simple on/off switch. The cell is integrating the strength of receptor binding, the duration of contact, the presence of co-stimulation, and the cytokines in the environment. That integration determines whether the T cell divides at all, how vigorously it does so, and which genes it turns on as it differentiates.

The dendritic cell's part in this is easy to overlook if one focuses only on the T cell. A dendritic cell is not a neutral screen displaying peptides; it actively curates what the T cell sees and how long it can be seen. Consider MHC class II, the platform read by CD4 T cells. Proteins taken up from outside the cell are broken into peptides in acidic compartments. MHC class II molecules are assembled with a placeholder fragment (a piece of a larger protein called invariant chain) that blocks premature peptide binding. In the peptide-loading compartment, that placeholder is trimmed down to a small remnant called CLIP. For the MHC to carry meaningful evidence, CLIP must be removed and replaced with peptides derived from the captured antigen. This exchange is not left to chance. A molecule called HLA-DM (in humans) acts as a catalyst and editor, helping dislodge CLIP and favoring peptides that form stable complexes with MHC class II. Stability matters because peptide—MHC complexes that persist longer on the surface offer a larger window for rare T cells to find them. One can think of HLA-DM as quietly raising the quality of the exhibit: fewer flimsy displays that fall apart quickly, more that endure long enough to be inspected.

MHC class I, the platform read by CD8 T cells, has its own quality-control machinery. Most peptides displayed by MHC class I come from proteins made inside the cell, which are chopped into fragments by the proteasome and transported into the endoplasmic reticulum by TAP (a transporter associated with antigen processing). There, MHC class I molecules are loaded with peptides in the company of chaperones. A protein called tapasin helps link MHC class I to TAP and encourages stable loading, though different MHC alleles vary in how much they depend on it. The details can feel technical, but the organizing theme is simple: antigen presentation is not a passive process of sticking fragments onto a surface. It is engineered to be both efficient and selective, because a naive T cell's first impressions can set the course of an entire immune response.

The helper versus killer split is often introduced as though it were predetermined: CD4 T cells help, CD8 T cells kill. The more interesting truth is that both are shaped by the context provided during priming. A CD8 T cell must acquire the machinery of cytotoxicity. It has to manufacture perforin, which can punch holes in target cell membranes, and granzymes, which can enter through those holes and trigger apoptosis, the programmed cell death pathway. It has to learn restraint as well as aggression, because indiscriminate killing would be catastrophic.

A CD4 T cell faces a different set of options. "Helper" is not one job description but a family of them. Some helpers, known as T_{H1} cells, produce cytokines that activate macrophages and support responses against intracellular pathogens. Others, T_{H2} cells, support responses against helminths and contribute to allergic inflammation. T_{H17} cells can mobilize neutrophils and are important in certain bacterial and fungal defenses, but can also contribute to inflammatory disease. T follicular helper (T_{FH}) cells are specialists in assisting B cells in germinal centers, enabling the production of high-affinity antibodies and long-lived memory. Regulatory T cells, often marked by the transcription factor FOXP3, function as peacekeepers, preventing excessive or misdirected responses.

These categories are real and useful, but they are not rigid boxes. Human immune responses often show mixed features, and the same infection can elicit different balances in different people. The underlying point is that helper T cells do not merely amplify immunity; they *choose its style*. That style is influenced by cytokines from dendritic cells and other innate cells, by the tissue environment, and by the microbial signals that initiated the response.

The split into CD4 and CD8 lineages also has a developmental origin. In the thymus, immature T cells are tested for their ability to recognize self MHC molecules weakly enough to be safe but strongly enough to be useful. Those that interact primarily with MHC class II tend to become CD4 cells; those that interact with MHC class I tend to become CD8 cells. By the time a naive T cell encounters a dendritic cell in a lymph node, it is already committed to reading one class of MHC better than the other. Yet the downstream differentiation—which type of helper, how cytotoxic, how long-lived—is molded during activation.

The immune system's need for both commanders and executioners becomes painfully clear in certain diseases. One of the most instructive historical examples is the AIDS epidemic. HIV targets CD4 T cells, using CD4 as part of its entry mechanism. As CD4 T cells decline, the immune system loses more than a single cell type; it loses coordination. Without adequate help, B cells produce poorer antibody responses, macrophages receive less activating instruction, and CD8 T cell responses can become less effective and less durable. Opportunistic infections that are normally controlled with ease begin to flourish. The clinical picture of AIDS is, in many ways, a demonstration of what helper T cells do when they are present: they make the rest of immunity competent.

A second example, different in flavor but equally clarifying, comes from transplantation. A transplanted kidney or heart is, immunologically, a dense package of foreign MHC molecules. T cells are exquisitely sensitive to these differences, and without immunosuppression many transplants would be rejected. Both CD4 and CD8 T cells can contribute. Helper T cells provide cytokines and co-stimulatory interactions that support rejection, while cytotoxic T cells can directly kill graft cells. Many immunosuppressive strategies aim, explicitly or implicitly, to interfere with T-cell activation and proliferation. The success of transplantation medicine is therefore an applied lesson in T-cell decision-making: change the signals at priming, and the entire immune response can be redirected or muted.

Under normal circumstances, the immune system uses this decision-making to tailor force. A viral infection in the respiratory tract is a good example because it requires both arms. Early in infection, innate responses limit spread, but viral replication inside epithelial cells demands a cytotoxic response. CD8 T cells expand, migrate to the infected tissue, and identify infected cells by scanning MHC class I peptides on their surface. When they recognize a matching peptide, they form a tight contact and deliver lethal granules, inducing apoptosis. This is not a dramatic explosion but a controlled demolition.

The target cell packages its contents to limit collateral inflammation, and phagocytes clear the remains.

At the same time, CD4 helper T cells are shaping the landscape. They support the activation and maintenance of CD8 responses. They help B cells produce neutralizing antibodies that prevent new infections of cells. They can recruit and activate innate cells to clear debris and prevent secondary bacterial invasion. The immune response becomes a coordinated campaign rather than a series of isolated skirmishes.

If that coordination sounds almost too purposeful, it is worth remembering that no cell possesses a master plan. The apparent strategy is an emergent property of rules: receptors, signals, and selection. A dendritic cell presenting antigen with high co-stimulation and particular cytokines creates a certain distribution of fates among responding T cells. Those fates then influence which other immune cells are recruited and how they behave. The system is deciding, but the decision is distributed.

The distinction between CD4 and CD8 is also a way of understanding why some pathogens are difficult to vaccinate against. Vaccines that excel at inducing antibodies may still struggle if the key to protection is a strong and appropriately targeted CD8 response. Conversely, vaccines that push strong cellular immunity may be less effective if neutralizing antibodies are the main protective factor. Vaccine design often involves the question: what kind of T-cell help is needed, and how can the antigen be presented so that the right lineage is engaged? Strategies such as using viral vectors, mRNA delivery, or specific adjuvants are, in part, methods for ensuring that antigens enter the MHC class I pathway (to prime CD8 cells) and that dendritic cells receive the innate signals that lead to robust co-stimulation.

There is also a moral hazard built into the system. Because T cells can be so powerful, the body maintains inhibitory pathways—molecular "brakes" that prevent runaway activation and limit damage to healthy tissue. These checkpoints are essential for preventing autoimmunity, but they can also be exploited by cancers, which may cultivate an immunosuppressive environment that keeps T cells exhausted or restrained. The modern success of checkpoint blockade therapies underscores, again, the centrality of T-cell decisions: restore a measure of permission, and dormant killing capacity can reemerge.

The story of commanders and executioners is therefore not simply a taxonomy of immune cells. It is a way of seeing how specificity becomes action. A peptide on an MHC molecule is inert information until it meets a receptor built to read it. Even then, recognition is not destiny. A T cell must be convinced that the recognition occurred in a context that justifies force, and it must be instructed about what kind of force is appropriate. CD4 T cells turn recognition into coordination: they amplify, specialize, and sustain the work of other cells. CD8 T cells turn recognition into direct intervention: they locate compromised cells and end them with precision.

The body pays a price for this precision. Killing infected cells can damage tissue, and intense helper-driven inflammation can make a person feel miserable: fever, aches, fatigue, the heavy fog of sickness. These are not side effects so much as the sensations of a system reallocating resources and accepting collateral costs in order to prevent a larger disaster. When the system works well, the misery is temporary and proportionate. When it fails—when help is absent, when killing is misdirected, when restraint is lost—the consequences can be lifelong.

At its best, the division of labor allows the immune system to behave with a kind of disciplined creativity. It can respond quickly, but also learn. It can coordinate large-scale defenses while still retaining the ability to eliminate a single corrupted cell. It can treat a virus as a problem of internal sabotage and a bacterium as a problem of external invasion, using different tools without needing different organs. A dendritic cell's careful presentation at the synapse begins the process, but it is the divergence of T cells into commanders and executioners that turns that information into a response with shape, pace, and intent.

3.3 Factories of Precision

A virus entering a cell is a private crime. Antibodies in the blood cannot reach it, and the immune system must rely on the intimate, face-to-face judgments of T cells. Many infections, however, depend on a more public stage. Bacteria spread through tissue fluid; toxins diffuse; viral particles travel between cells; parasites move in the bloodstream. In these situations, it is not enough to eliminate infected cells one by one. The immune system needs a way to patrol open space, to bind threats before they dock, to mark them for disposal, to neutralize their tools. That is the special talent of antibodies.

An antibody is a protein made by B cells, and in the right circumstances it behaves like a guided missile: it seeks a particular molecular feature and binds with such tenacity that the target's options narrow. A toxin can no longer attach to its receptor. A virus can no longer fuse with a cell membrane. A bacterium becomes coated in a signal that says *eat this*. Yet the really astonishing part is not what antibodies can do, but how quickly and in what numbers the body can make them once a B cell has been properly activated.

A single plasma cell—a B cell transformed into a dedicated antibody secretor—can release thousands of antibody molecules every second. Multiply that by millions of plasma cells during a robust immune response, and the scale becomes almost industrial. The lymph node and spleen, which began as meeting places, become manufacturing districts. The bloodstream becomes a distribution network. Antibodies saturate tissues and mucosal surfaces, turning the body's fluids into an active medium rather than an inert background.

This manufacturing capacity would be meaningless without specificity. Making a trillion copies of the wrong product is not a triumph; it is a waste. B cells must be able to produce antibodies that bind a pathogen and ignore everything else, including the body's own molecules and the harmless clutter of food and friendly microbes. How does the immune system build a repertoire of antibodies broad enough to anticipate the unknown?

The answer begins long before infection. In the bone marrow, developing B cells perform a kind of controlled genetic vandalism. They cut and paste segments of DNA to assemble the gene that encodes the antibody's variable region, the part that will eventually touch the antigen. This process is called V(D)J recombination because it uses three classes of gene segments: V (variable), D (diversity), and J (joining). Enzymes called RAG1 and RAG2 guide the cutting at specific DNA motifs (recombination signal sequences), enforcing a rule about how segments can be paired, and the cell repairs the breaks using a pathway called non-homologous end joining. The product is a new gene sequence that did not exist in the inherited genome, and therefore a new binding shape that has never been tested by evolution.

It is hard to convey how audacious this is. Most cells treat double-strand breaks in DNA as emergencies. A B cell deliberately creates them as part of normal development. The benefit is enormous diversity: a population of B cells, each with a different antibody receptor, able to recognize an enormous space of possible molecular surfaces. The cost is risk. If recombination goes wrong—if breaks occur at the wrong site or are joined incorrectly—chromosomes can rearrange, and the cell can be pushed toward malignancy. Certain leukemias and lymphomas are, at their root, accidents of the very machinery that makes adaptive immunity possible. The immune system's precision is built on a willingness to tolerate calculated danger.

A mature B cell leaves the bone marrow carrying its antibody as a receptor on its surface. At this stage, the antibody is not yet a soluble weapon; it is a sensor. The B cell uses it to sample the world. When antigen binds, the B cell internalizes the antigen, processes it, and displays peptides from it on MHC class II molecules. That last step brings the earlier logic of T-cell help into the B cell story. A B cell is not fully activated just because its receptor has bound something. It must usually receive approval from a helper T cell, particularly a T_{FH} cell, which recognizes the B cell's peptide–MHC display and provides crucial co-stimulation and cytokines.

This two-key system—B cell recognition plus T cell help—is another safety interlock. It reduces the chance that a B cell reacting to a self molecule will be allowed to expand and secrete damaging antibodies. It also ensures that antibody responses are not merely sticky but relevant: T cells have already been vetted through thymic selection and require their own activation conditions. The immune system asks for concordance between two independent detectors before committing to mass production.

When that concordance is achieved, B cells can take two major paths. Some become short-lived plasma cells quickly, providing an early wave of antibodies that can begin containment. Others enter a more elaborate structure called a germinal center, which forms within follicles of lymph nodes and the spleen. A germinal center is where antibodies are refined. It is one of the immune system's most impressive pieces of evolutionary engineering: a microenvironment that deliberately accelerates mutation in a specific genomic region and then selects the best outcomes.

The germinal center is not a uniform blob. Under the microscope it shows distinct zones, often described as a dark zone and a light zone. In the dark zone, B cells proliferate rapidly. They are called centroblasts there, and they engage in somatic hypermutation: a process that introduces point mutations into the variable region of the antibody gene. The initiating enzyme is AID (activation-induced cytidine deaminase), which converts cytidine to uracil in DNA at immunoglobulin loci. That chemical change is a small vandalism, but it invites repair pathways that are intentionally error-prone in this setting. Mutations accumulate. Many will weaken binding; some will destroy it; a few will improve it.

In the light zone, B cells (now often called centrocytes) compete for antigen. The antigen is not usually floating freely. It is displayed in immune complexes and presented by follicular dendritic cells—cells that are not the same as the dendritic cells that prime T cells, but which serve as long-term display platforms within follicles. B cells with higher-affinity receptors can capture more antigen, internalize more, and present more peptide on MHC class II. This improves their chances of receiving help from T_{FH} cells. Help is not merely encouragement; it is a survival signal. Without it, many B cells die. With it, selected B cells can return to the dark zone for further rounds of mutation or can differentiate into plasma cells and memory B cells.

A germinal center, then, is a Darwinian process running inside the body on a compressed timescale. Variation is generated by mutation, and selection is enforced by competition for a limited resource (antigen and T-cell help). The output is an antibody response that often improves over weeks. Anyone who has watched serology results change after infection—antibodies rising in amount and often in quality—has seen the population-level signature of this microscopic tournament.

There is another kind of refinement that happens in activated B cells, also driven by AID: class-switch recombination. Early in a response, B cells often produce IgM, an antibody form that is effective at certain tasks, especially early complement activation, but which is not always optimal for long-term tissue defense. Through class switching, the B cell changes the constant region of its antibody heavy chain, altering the antibody's effector properties without changing its antigen specificity. IgG subclasses can diffuse well into tissues and persist in blood. IgA is specialized for mucosal surfaces, including the gut and respiratory tract, and can be transported across epithelia. IgE binds strongly to mast cells and basophils and is potent in defense against parasites, but also central to allergic disease. The choice of class is influenced by cytokines from helper T cells and the tissue environment, which means that the immune system does not only ask *what* to bind; it asks *how* to deploy what it binds.

The phrase "antibody production" can make this sound like a simple faucet turned on. In reality, plasma cells are among the most extreme secretory machines in the body. Their internal architecture changes to support this role: the endoplasmic reticulum expands, the Golgi apparatus becomes prominent, protein-folding and quality-control pathways run at high capacity. The cell's energy demands rise. A plasma cell's identity is, in a real sense, a bet that the immune system is making: invest heavily in a single product line because the threat is urgent enough and the evidence is solid enough.

The payoff is not merely quantity, but a kind of distributed surveillance. Antibodies do not have to find a pathogen in a specific lymph node. They circulate widely. They seep into tissues. They line mucosal surfaces. They can bind pathogens before those pathogens find their preferred entry points. And when they bind, they recruit other forms of help. The antibody's constant region is a handle recognized by Fc receptors on phagocytes and by complement proteins. A coated bacterium is easier for a macrophage to ingest. A virus covered in antibodies is more likely to be trapped and cleared. Antibodies also neutralize, a term that sounds gentle but can be decisive: blocking a receptor-binding site on a virus can turn a dangerous particle into inert debris.

A concrete example makes the logistical scale emotionally real: diphtheria. Before widespread vaccination, diphtheria was a feared childhood disease, not primarily because the bacterium invaded deeply, but because it produced a potent toxin. The toxin could cause cardiac and neurological damage, and the gray pseudomembrane in the throat could suffocate. The medical breakthrough that began to tame diphtheria involved antitoxin serum: antibodies raised in animals and transferred to patients. Emil von Behring's work in the 1890s showed that blood serum could carry protective factors, earning him the first Nobel Prize in Physiology or Medicine in 1901. The concept was simple and profound: immunity could be *moved* from one body to another through antibodies. This was not yet a vaccine, but it revealed the power of humoral immunity and the fact that the bloodstream could be stocked with a targeted neutralizing reagent.

That history also highlights a limitation. Passive antibody transfer can save a life in the moment, but it does not always build lasting memory. For enduring protection, the body needs its own B cells and plasma cells to produce antibodies over time. Some plasma cells are short-lived, built for a burst of output. Others become long-lived and settle in supportive niches, particularly in the bone marrow, where they can secrete antibodies for years. Memory B cells, for their part, do not secrete large amounts of antibody immediately. They are a reserve force: poised, experienced cells that can rapidly re-enter germinal centers or differentiate into plasma cells upon re-exposure to antigen. The memory response is often faster and higher affinity, because the starting pool has been pre-selected and refined.

The elegance of this system is matched by its capacity for mischief.

The same processes that generate high-affinity antibodies against pathogens can generate high-affinity antibodies against the body's own tissues if tolerance fails. Autoantibodies can cause disease directly by binding receptors, forming immune complexes that inflame tissues, or marking cells for destruction. Allergies represent another misdirection: IgE responses that are harmless in their original evolutionary context (parasite defense) become pathological when directed at pollen, foods, or dander. In both cases, the antibody factory is not broken; it is working too well for the wrong target.

Even when properly targeted, antibodies are not omnipotent. Some pathogens evolve rapidly, changing their surface proteins to evade binding. Influenza is an obvious example: antigenic drift means that the antibodies refined one year may fit poorly the next. HIV presents a harder challenge, cloaking critical surfaces with sugars and mutating key regions, forcing the immune system into a chase where its refinement machinery is constantly being outrun. Yet even in these cases, the existence of germinal centers and affinity maturation shapes outcomes. Broadly neutralizing antibodies, when they arise, are often products of long germinal center evolution, with extensive mutation and selection.

The B cell response also reveals something philosophically interesting about immune "decision-making." A germinal center does not plan. It experiments. It throws variation at a problem and rewards what works under the constraints of the body's internal environment. The immune system learns, but it learns by generating options and pruning. The fact that this happens in DNA—that the cell rewrites the very sequence that encodes its receptor—gives immune memory a physical permanence that feels almost like experience etched into tissue.

When a person recovers from an infection and remains protected, it is tempting to imagine a single durable token stored somewhere: a memory cell labeled with the pathogen's name. The reality is more like a supply chain with redundancies. Memory B cells circulate and reside in tissues, ready to restart the refinement process. Long-lived plasma cells quietly secrete antibody, maintaining a baseline of protection. The antibody molecules themselves, floating in blood and bathing mucosal surfaces, are the most immediate manifestation of past experience: a chemical record that says, *this has been seen before, and the body has already built the tools to bind it.*

In the everyday world, this machinery is largely invisible until it fails or until it is intentionally manipulated. A booster shot works because memory B cells and residual germinal center potential can be recruited again, sharpening affinity and replenishing antibody levels. Monoclonal antibody therapies work because scientists can now do, in the laboratory, what the immune system does in the body: isolate a single successful antibody blueprint and manufacture it at scale. In both cases, the goal is the same as the goal of a plasma cell: to flood a vulnerable space with molecules shaped to fit a particular target.

The image of plasma cells as factories is not just metaphorical. They are factories in the way that matters: they convert information into material. The information is the shape of an antigen, refined by mutation and selection and validated by T cell help. The material is protein poured out into the bloodstream in staggering quantities, ready to bind, neutralize, and recruit other forces. When that conversion succeeds, the immune system is no longer merely reacting; it has altered the body's baseline chemistry in a way that makes a future encounter with the same threat feel, to the pathogen, like trying to move through a room already filled with glue.

Chapter 4

Evolution in Real-Time

Natural selection is usually measured in eons, but within our lymph nodes, it occurs on a timescale of days. We enter the chaotic, high-stakes environment of the germinal center to witness the rapid genetic engineering that refines our biological weaponry. From this microscopic Darwinian struggle emerges a tiered defensive network-from circulating patrols to stationary guards-capable of remembering a specific enemy for a lifetime.

4.1 Survival of the Fittest Molecules

A lymph node can feel like an unremarkable object until you remember what it is built for: traffic. Lymph, the clear fluid that drains from tissues, carries scraps of whatever has been happening out at the body's borders—a splinter's bacteria, a virus from the throat, proteins from pollen, debris from ordinary wear and tear. The lymph node is a checkpoint where immune cells inspect those scraps and decide which ones matter. Most of the time, nothing dramatic follows. But when a genuine threat is present, certain regions of the node flare into intense activity, forming pale, transient structures called *germinal centers*. Under the microscope they look like islands. Functionally, they are more like laboratories that run on impatience.

What happens inside a germinal center is one of the most daring things biology does in peacetime: it deliberately damages its own DNA in order to improve a weapon. The weapon is the antibody molecule, and the cells producing it are B cells. A B cell carries, on its surface, a version of the antibody it is capable of making; this surface-bound form is called the B cell receptor (BCR). It is the B cell's sensor and its hook. When a B cell's receptor latches onto a piece of a microbe—an *antigen*—the B cell is not simply "turned on." It is recruited into a tournament.

The tournament begins with a small number of "founder" B cells that have already recognized the invader well enough to be worth investing in. They enter the germinal center and start dividing rapidly. Then comes the dangerous trick: *somatic hypermutation*. "Somatic" means it happens in the body's cells, not in eggs or sperm; it is not inherited by children. "Hypermutation" means mutation turned up to a rate that would be catastrophic almost anywhere else. The mutations are targeted mainly to the small stretches of DNA that encode the antigen-binding tip of the antibody. In effect, the immune system takes an already serviceable binding site and starts generating variants of it—not one or two, but a cloud.

This is not the slow, population-level evolution described in textbooks about finches. It is evolution compressed into days, operating within one person's lymph node. Variation is generated on purpose. Selection is ruthless. And the trait under selection is molecular: the tightness and usefulness of an antibody's grip.

It helps to be precise about what "better" means here. A high-affinity antibody is one that binds its target antigen strongly—like a key that fits a lock with little wobble. But the germinal center does not directly "measure" affinity with a tiny force gauge. Instead, it uses consequences. A B cell with a receptor that binds more tightly is more likely to grab antigen from the surfaces where it is displayed. That captured antigen is pulled inside the B cell, chopped into peptides, and displayed on the B cell's surface on molecules called MHC class II. Those peptide–MHC II displays are a kind of billboard that says, in biochemical language, *this is what I found.* The more antigen a B cell acquires, the more convincing its billboard tends to be.

The judges in this tournament are specialized T cells called *T follicular helper* (Tfh) cells. A Tfh cell inspects a B cell's peptide–MHC II display and, if satisfied, delivers help: signals that prevent the B cell from dying, encourage it to divide again, and steer its fate. A B cell that fails to win enough Tfh attention undergoes apoptosis, a programmed cell death that is tidy but final. The germinal center is thus a place where survival depends on persuasion. A B cell must not only bind antigen; it must convert binding into a strong enough presentation to earn T cell support.

Germinal centers are organized to make this process iterative. They contain two main regions, often described as zones: a *dark zone*, dense with dividing B cells, and a *light zone*, where selection happens. The names come from staining patterns on microscope slides, but they capture the logic: in one area, B cells proliferate and mutate; in the other, they compete and are judged. Cells shuttle between these zones in cycles. Each cycle is a generation in miniature: variation, competition, selection, and then reproduction of the winners.

Right after somatic hypermutation is introduced, it is useful to see the overall choreography as a loop:

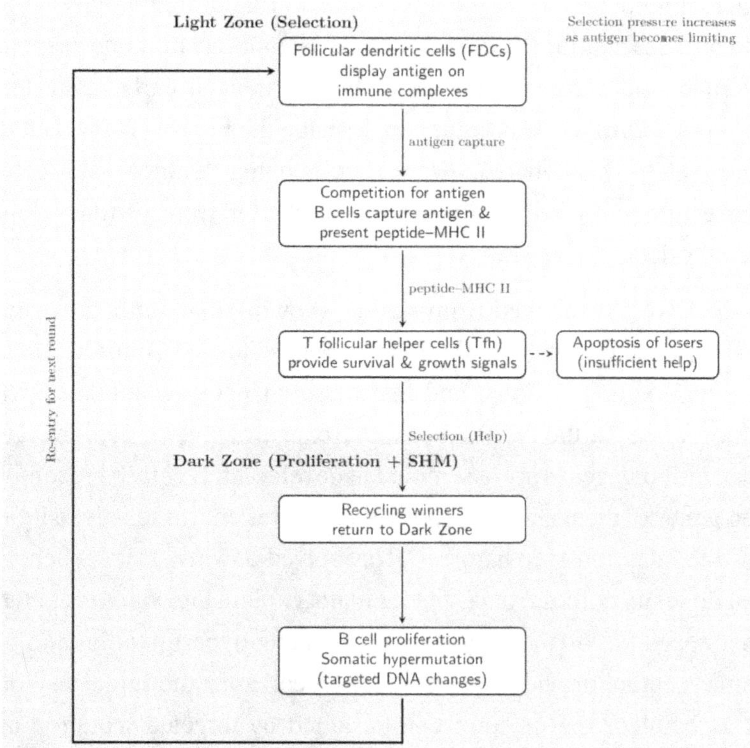

Even in this simplified diagram, two features matter for the story. First, selection is not only about binding; it is about *winning help*. Second, the cycle repeats, meaning that small advantages compound. If a particular mutation gives a B cell a slightly better grip, that B cell tends to present more antigen, receive more Tfh help, undergo more divisions, and generate more descendants—some of which will carry additional mutations. The result can look like a smooth improvement in antibody quality over time, but underneath that smooth curve is a constant culling.

The most vivid evidence that something like this was happening came from classic experiments that tracked how antibodies changed after immunization. In the 1960s and 1970s, immunologists could already see that antibodies produced later in an immune response often bound antigen better than those produced early. The term *affinity maturation* was coined to describe the improvement. But the mechanism was debated. Were B cells being "trained" in a gentle sense, gradually adjusting like muscles? Or were distinct clones being selected, Darwin-style?

A decisive clue arrived from studies showing that antibodies could differ by small changes in their amino acid sequences, clustered in the antigen-binding regions, and that these changes accumulated with time. Then, in the 1970s, Susumu Tonegawa's work demonstrated that antibody genes are assembled in developing B cells by recombining gene segments, a mechanism that creates enormous diversity before an infection even happens. That solved one mystery—where initial diversity comes from—but it did not explain the *ongoing* changes that appeared after immunization. Somatic hypermutation and germinal centers provided the missing piece: after the initial receptor is assembled, the receptor can be tuned by targeted mutation and selection.

A concrete example makes the stakes clear. Consider the difference between meeting a pathogen for the first time and meeting it again. Early antibodies tend to bind moderately well. They can still be lifesaving, especially when made quickly and in large amounts. But microbes multiply. A virus particle that escapes one antibody can seed more. The immune system benefits immensely from antibodies that bind not just "well enough" but exceptionally well: antibodies that cling even when the pathogen is scarce, antibodies that neutralize with fewer molecules, antibodies that outcompete the pathogen's own attempts to attach to our cells. Germinal centers exist because "good" is not good enough when the opponent evolves too.

The germinal center's method is risky because mutation is risky. A single letter change in DNA can improve binding, but it can also ruin it. Most changes are neutral or harmful. Many mutated B cells will produce receptors that bind antigen less well than before, or not at all. Some will produce receptors that are unstable or misfolded. The germinal center has no sentimentality about these failures. If a B cell cannot gather and present enough antigen to obtain Tfh help, it dies. Under a microscope, tingible body macrophages—cells that clean up apoptotic debris—are a prominent feature of germinal centers. Their presence is a quiet signature of the scale of waste. The immune system spends cells the way a factory spends sparks: constantly, as part of the mechanism.

Selection pressure can intensify as the response progresses. One reason is that the very antibodies being produced feed back into the system. As antibodies accumulate, they bind antigen and form immune complexes that can be displayed on follicular dendritic cells (despite their name, these are not the same as the dendritic cells that activate T cells in the first place). Antibody binding can mask certain parts of the antigen and leave others exposed, changing what is easiest to grab. This means that the "test" B cells face is not static. The target is the same pathogen, but the antigen is increasingly decorated with the immune system's own products. Late in the response, when free antigen is scarce, only B cells with receptors that can compete under these conditions will thrive.

Another reason selection becomes more stringent is simply crowding. Tfh help is not unlimited. A germinal center contains many B cells and fewer Tfh cells, and their interactions are brief. Help becomes a limiting resource. That limitation is not a flaw; it is part of the design. Scarcity forces discrimination. If help were abundant, mediocre B cells would survive, and the system would settle for adequacy. By making help hard to obtain, the germinal center amplifies small differences in performance.

This is also where the immune system reveals a deeper kind of "decision-making," one that looks almost strategic even though it is built from molecules and contact-dependent signals. Tfh cells do not merely sprinkle generic encouragement. The quality and quantity of help they deliver influences whether a B cell returns for another round of mutation, exits to become an antibody-secreting plasma cell, or leaves as a memory B cell. These fates matter. A plasma cell is a factory: it produces large amounts of antibody, often at the cost of flexibility. A memory B cell is a reserve: quieter, more mobile, able to re-enter responses later. In a germinal center, the immune system is effectively budgeting: how many resources to spend now on immediate antibody output, and how many to invest in a future that may or may not arrive.

The popular image of evolution is a slow grind, but the germinal center's pace is closer to a frantic rehearsal where the cast changes every night. B cells can divide quickly, and each division is a chance for mutation in the receptor genes. The result is a shifting population in which "lineages" of related B cells can be traced, like family trees, through their shared mutations. Some lineages flourish; others end abruptly. If you could watch the process in real time—cells moving, binding, competing, dying—it would look less like a calm optimization routine and more like a high-turnover ecosystem.

And yet the outcome can be remarkably precise. After a successful response, the body can produce antibodies that recognize a pathogen with astonishing specificity, distinguishing between proteins that differ by tiny structural features. This specificity is not a philosophical ideal; it is practical. It allows antibodies to neutralize a virus without sticking indiscriminately to the body's own proteins. It allows an immune response to be potent without being permanently inflamed. The germinal center is one of the places where potency and restraint are negotiated.

There is an emotional angle to this biology that rarely makes it into diagrams. A person who receives a vaccine is, in a quiet way, asking their immune system to run this tournament under controlled conditions. The antigen is presented without the full dangers of disease, giving the immune system time to refine its tools. In the days and weeks after vaccination, germinal centers can persist, continuing to generate and test variants. What feels like a moment—a shot in the arm—is followed by an extended period of invisible experimentation in lymph nodes, as if the body is rehearsing for a crisis it hopes never comes.

The germinal center is also a reminder that biological "learning" does not require a brain. Nothing in the lymph node understands a pathogen in the human sense. There is no plan, no foresight, no internal narrative. There are only cells that bind or fail to bind, present or fail to present, receive help or fail to receive it. But from that accumulation of local interactions emerges something that looks uncannily like improvement with experience. The antibodies made later can bind more tightly than the antibodies made early. A response can become more focused. A memory can be formed that is better than the first attempt.

If you hold onto one image, let it be this: a microscopic arena where molecules are treated as contestants, and survival is granted to those that can perform one task slightly better than their neighbors. The immune system wagers that it can afford the casualties. It wagers that, given enough variation and enough culling, it can sculpt a binding site that fits the enemy so well that the next encounter will feel, to the pathogen, like trying to swim while wrapped in glue.

4.2 The Sentinels That Remain

After the fever breaks and the swollen lymph nodes settle down, it is tempting to imagine the immune system returning to a blank slate. Many of its most dramatic cells do vanish. The short-lived infantry of an acute response is designed to surge and then die back, partly to save resources and partly to reduce the risk of collateral damage. Yet immunity that has learned something rarely lets that lesson evaporate. It keeps a record, and that record is not stored in a single place or a single form. It is distributed among cells that behave less like a standing army than like a security system with different kinds of sensors: some roam, some wait at chokepoints, and some take up residence so thoroughly that they begin to look like part of the tissue itself.

A useful starting point is to separate two ideas that everyday language tends to blur. One idea is *memory as information*: the immune system has, after an infection or vaccination, a larger number of cells that can recognize a particular antigen, and many of those cells carry receptors tuned by selection. The other idea is *memory as geography*: where those experienced cells choose to live. Geography matters because infection is not a polite visitor. It enters through particular doors: the nose and lungs, the gut, the skin, the genital tract. A memory cell that lives in the wrong place may still be valuable, but it will arrive late, and sometimes late is the same as never.

T cells make this geographic logic especially clear. T cells are the immune system's specialists in recognizing infected cells and coordinating responses. Their receptors (T cell receptors, or TCRs) do not bind free-floating antigen the way antibodies do. Instead, T cells recognize peptide fragments displayed on MHC molecules by other cells. In a first encounter with a pathogen, naive T cells are activated in lymph nodes after dendritic cells ferry in antigen from the infected tissue. Many days can pass between infection and the moment a useful number of effector T cells have been produced and sent out. That delay is one reason people feel sick: the pathogen gets a head start.

Once a T cell has been activated and expanded, some of its descendants become memory T cells, long-lived cells poised to respond again. But memory T cells are not all the same kind of poised. Immunologists often distinguish at least two broad circulating populations. *Central memory* T cells tend to circulate between blood and lymph nodes. They are good at proliferating when they see antigen again, generating a large secondary wave of response. They resemble a reserve force that can be mobilized quickly once the alarm reaches headquarters. *Effector memory* T cells circulate more through blood and peripheral tissues and can respond with less delay, often carrying pre-formed tools for immediate action. They resemble patrols: not stationed in one room, but in the neighborhood.

That still leaves a third mode of existence that, once discovered, changed how immunologists thought about protection. Some memory T cells do not circulate in any ordinary sense. They settle into tissues—skin, lung, gut, reproductive tract, even certain niches in the brain—and they stay. These are called *tissue-resident memory* T cells, often abbreviated TRM. Their residence is not an accident of being temporarily stuck. It is an identity. They express molecules that help them remain anchored and that reduce their tendency to re-enter the bloodstream. They adapt to the local environment, drawing on signals from the tissue and, in return, shaping the tissue's readiness for danger.

If this sounds like a minor rearrangement, it is worth pausing on why it matters. Many infections begin locally and spread. Influenza begins in the respiratory tract. Herpes simplex establishes itself in skin or mucosa and then in nerves. Norovirus begins in the gut. In those early hours, a resident memory cell can act before a pathogen has multiplied enough to cause obvious symptoms. It can be the difference between a brief skirmish and a city-wide fire.

There is also something psychologically satisfying about the idea: that the body remembers not only *what* attacked but *where* it attacked. The skin is not simply a wall; it is an organ with a history. After a bad sunburn or a severe dermatitis flare, the skin can behave differently for months. Some of that change is due to altered barriers and local nerves. Some is immune. TRM provide a cellular mechanism for that sense of local memory.

A concrete historical turning point came from experiments that tested whether immune protection could persist in a tissue even when circulating T cells were excluded. Techniques like parabiosis—surgically joining two mice so that they share a bloodstream—were used to distinguish truly resident cells from cells that simply pass through. If a memory T cell population in the skin remained largely in the original mouse even after months of shared circulation, it suggested residence rather than constant replenishment. These studies, along with careful tracking of cells and their surface markers, built the case that many tissues maintain their own local cadre of experienced T cells.

The result is a more realistic picture of immune surveillance. The immune system does not rely on a single pool of memory cells circulating like lifeguards endlessly circling a lake. It uses layered strategies. Some memory is centralized, ready to expand. Some memory is mobile, ready to act. Some memory is embedded, ready to intercept.

The same geographic questions apply to B cells, but in a different way. B cells' most famous product is antibody, and antibodies are wonderfully portable. A protective antibody circulating in the blood can neutralize a virus before it enters cells. Antibodies can leak into tissues, coat mucosal surfaces, and be transported across epithelia. So one might think B cell memory is mostly a matter of maintaining antibody levels and keeping a few memory B cells in circulation. That is true, but incomplete.

One overlooked form of "sentinel" is the long-lived plasma cell. Plasma cells are the final, factory-like form of activated B cells, specialized for producing antibody in enormous quantities. Many plasma cells generated early in an immune response are short-lived, dying after days or weeks. But a subset becomes long-lived, taking up residence primarily in the bone marrow and continuing to secrete antibody for years, sometimes decades. They are memory in the most literal biochemical sense: they keep a molecule in circulation that can act immediately, without waiting for any cell to divide.

These long-lived plasma cells are not immortal. They survive because they occupy supportive niches—microenvironments in the bone marrow that provide contact signals and soluble factors that prevent apoptosis and sustain antibody secretion. The niche is not a single "plasma cell hotel" with identical rooms. It is more like a patchwork of supportive neighborhoods shaped by stromal cells, blood vessels, and other hematopoietic cells. Mouse studies have suggested that eosinophils, better known for their roles in allergy and parasitic infections, can contribute survival signals such as APRIL and IL-6 to plasma cells in the marrow. More recent work in human systems has emphasized the importance of stromal contact and particular survival pathways inside the plasma cell, including PI3K and canonical NF-κB signaling. The details differ across models and species, but the overarching point is steady: durable antibody production is not merely a property of the plasma cell itself. It is a partnership with a place.

That partnership has clinical consequences that can be surprising. In some autoimmune diseases, pathogenic antibodies can persist even after treatments that deplete circulating B cells. People can reasonably assume that "removing B cells" removes antibodies. Often it does reduce new antibody generation. But long-lived plasma cells in their marrow niches may remain largely untouched, continuing to secrete the very antibodies driving disease. This is one reason autoimmune conditions can be stubborn: the immune system's memory is not only cellular; it is architectural.

Now return to the idea of tissue residence, but for B cells. For a long time, "memory B cell" conjured a circulating cell that could be found in blood or secondary lymphoid organs and that would quickly proliferate on re-exposure. That remains central. Yet it turns out that B cells, too, can form locally persistent populations in non-lymphoid tissues. Tissue-resident memory B cells (often abbreviated BRM) have been described in places like the lung after respiratory infections. They are not merely passing through on their way to a lymph node. They can remain and contribute to local protection.

To see why lung BRM might matter, imagine the early hours of re-infection with influenza. The airways are a fast-moving interface, constantly exposed to the outside world and constantly swept clean by mucus and cilia. A virus that gains a foothold can spread quickly along the epithelium. A circulating memory B cell that needs to be reactivated in a lymph node, proliferate, and differentiate into plasma cells will not help much in those first crucial hours. A pre-existing antibody level may help; so can TRM. But a local B cell population offers another advantage: it can generate antibody *right there*, rapidly, potentially focusing production on the respiratory tract where it is needed most.

In mouse models, local memory B cells in the lung have been shown to respond rapidly on re-exposure, differentiating into antibody-secreting cells and helping to control infection. The exact timing depends on the model and the pathogen, but the conceptual shift is robust: the tissue itself can host a memory module capable of making antibodies without waiting for distant coordination. The lung is not only defended by circulating factors; it can carry its own immunological recollections.

Residence, however, has a double edge. A permanently stationed guard can protect, but it can also overreact. Barrier tissues are the places where immune mistakes are felt most acutely: asthma in the lungs, inflammatory bowel disease in the gut, dermatitis in the skin. Tissue-resident memory cells, both T and B, are increasingly implicated in the persistence of such conditions. The same features that make them valuable against reinfection—longevity, rapid responsiveness, a local foothold—can also make inflammation hard to extinguish. The immune system, in other words, pays for speed with the risk of stubbornness.

Geography also helps explain an everyday mystery: why immunity can be strong in one compartment and weaker in another. An injected vaccine tends to elicit strong systemic immunity—antibodies in blood, memory cells that circulate. But mucosal immunity in the nose or gut can be harder to generate with an injection alone, because the immune system's "postal codes" matter. Cells activated in a particular tissue environment acquire homing molecules that guide them back to related sites. Likewise, local antigen exposure is often especially good at establishing tissue-resident populations. This is not a failure of vaccines; it is a clue to how specific the immune system's deployment rules can be.

What, then, does it mean for immune memory to "patrol the blood" versus "take up permanent residence"? It means that the body hedges. Central memory T cells and many memory B cells circulate and can be sampled in blood tests, which is convenient for clinicians and researchers. But those blood samples are, in a sense, an airport lounge: busy, informative, and not where most people live. Some of the most consequential immune memory cells are in tissues, quietly doing the unglamorous work of being ready.

A concrete illustration comes from the skin. The skin covers about two square meters in an adult and is constantly exposed to minor injuries and microbes. Memory T cells that reside in the skin can respond to familiar threats quickly and can also help recruit other immune cells. This is protective in the face of recurrent exposure. But it also helps explain why certain skin diseases flare in the same locations. A patch of skin that has hosted a strong immune response may be left with a durable population of resident memory cells primed to respond again. Even if the original trigger is gone, a new irritation can be interpreted through the lens of that local history.

In the lungs, the story becomes even more personal. Respiratory viruses are among the most common infections humans experience, and their outcomes range from a mild cold to pneumonia. The difference between those outcomes is shaped by age, underlying health, and the pathogen itself, but also by the positioning of immune memory. A person with strong neutralizing antibodies circulating may block infection early. A person with robust lung TRM may clear infected cells before the virus spreads deeply. A person with neither may still recover, but only after the slower machinery of activation and expansion has time to spin up. When people speak of "having a good immune system," they are often unknowingly speaking about these layers of readiness and the time they buy.

The immune system's choice of where to station memory is, in the end, a choice about time. Circulating memory is flexible and wide-ranging, capable of covering many tissues but requiring signals and travel. Resident memory is narrow but immediate, trading breadth for speed. Long-lived plasma cells are neither roaming nor local guards in the usual sense; they are infrastructure, providing a baseline chemical defense that is always on. Together, these strategies turn the immune system's past experiences into present capability, not as a single stored file but as a set of living deployments.

A final nuance makes the picture more human. The immune system is not a solitary entity; it is embedded in a body with habits and environments. The tissues most exposed in a person's life—the lungs of a city dweller, the skin of someone who works outdoors, the gut of someone whose diet changes dramatically, the airways of someone exposed to smoke—are the tissues where immune encounters are most frequent. Over years, the map of resident memory cells becomes, in part, a map of a life lived. The sentinels that remain are not only guardians against the next infection; they are a record of where the world has touched the body, and where the body has decided it cannot afford to be surprised again.

4.3 The Long Wait

The most unsettling feature of immune memory is not its speed, but its patience. A person can recover from measles and carry protection for a lifetime. Someone vaccinated against tetanus may be protected for years and then, without warning, that protection begins to thin and requires renewal. The immune system's long-term vigilance is not a single stored object, like a photograph kept in a drawer. It is a living arrangement that must be maintained: cells that must not die, molecules that must not decay too quickly, and a body that must continue to tolerate these quiet sentinels without treating them as clutter.

The time scales are bizarre. In the space of a week, a rare lymphocyte can divide into an army; in the space of decades, some of its descendants can remain poised to act again. "Poised" does not mean constantly active. Constant activity would be exhausting and dangerous. Most memory cells spend most of their lives in a state closer to watchful idling: low-level signaling, careful metabolism, the cellular equivalent of keeping the engine warm enough to start.

The problem is easiest to appreciate if you imagine what the immune system is up against. Cells are not hard drives. Proteins wear out. DNA is damaged and repaired. Organisms age. Even if a memory cell could, in principle, sit unchanged for thirty years, its environment would change around it. Nutrients vary, hormones fluctuate, infections come and go, inflammation rises and falls. The immune system must keep memory functional without keeping it inflamed.

One answer is that memory is spread across different kinds of long-lived entities. Long-lived plasma cells in the bone marrow provide antibodies continuously, as if a faucet has been left slightly open. Circulating and tissue-resident memory T cells provide rapid cellular responses. Memory B cells provide the capacity to generate new plasma cells and new rounds of antibody improvement. None of these, alone, fully explains decades of protection. Together, they create redundancy: if antibodies decline, memory B cells can replenish them; if a virus slips past antibodies, memory T cells can limit damage; if one tissue is poorly defended, another layer can compensate.

Even with redundancy, persistence requires maintenance rules, and those rules are surprisingly subtle. Memory cells do not merely "survive." They are actively kept alive by signals that are small in amplitude but steady in effect. T cells, for example, depend on cytokines (immune signaling proteins) such as IL-7 and IL-15 for homeostatic survival and slow turnover. These cytokines do not tell a memory T cell to attack; they tell it, in essence, *stay ready.* B cells receive analogous survival cues through other receptors and niches. These signals are part of a broader principle: immune memory is not a permanent tattoo but a relationship that the body continually renegotiates.

A second rule is that memory is not purely static; it can be renewed by occasional division without losing identity. Many memory T cells and memory B cells undergo slow homeostatic proliferation. This is not the explosive proliferation of an acute response. It is a gentle replacement of cells that die, keeping numbers roughly stable. If you picture a long-running theater production, it is less like a single cast member staying on stage forever and more like a company that maintains the same play over many seasons, replacing individuals as needed while preserving the script.

That renewal raises a problem of fidelity: how do cells divide without drifting into irrelevance? The immune system's solution is to keep the receptor genes stable in memory lymphocytes. A memory T cell does not keep mutating its TCR. A memory B cell *can* acquire additional mutations if it re-enters a germinal center, but it does not typically continue mutating in the background. In between encounters, memory is conserved by preserving the receptor and preserving the cell's readiness program.

Readiness has a metabolic dimension that is often missed in popular accounts. Effector cells of an acute response are metabolically extravagant. They grow, divide rapidly, and produce large amounts of protein. They rely heavily on glycolysis, a fast way of generating energy and building blocks. Memory cells, by contrast, tend to adopt a more economical metabolism, relying more on oxidative phosphorylation and fatty acid oxidation, maintaining mitochondria in good working order. This is not just housekeeping. Metabolism shapes what a cell can do quickly. A memory cell that is metabolically "fit" can switch gears fast, ramping up protein synthesis and division when needed. A cell that has limped along for years without proper maintenance cannot.

Quiet vigilance also requires restraint in signaling. Immune activation pathways can be self-amplifying: once a cell begins producing certain cytokines or transcription factors, it can push itself further into activation. Memory cells keep many of these pathways partially primed but not fully engaged. They often maintain chromatin (the packaging of DNA) in a state that makes key genes easier to turn on quickly. The genes are not being shouted; they are being kept within easy reach.

These cellular strategies address the "how" of waiting. The "why" of waiting is revealed by reinfection, when memory turns time into survival. Yet the most fascinating twist is that waiting can be productive even *before* reinfection happens. The immune system is not always done evolving when symptoms end. Germinal centers, those evolutionary pressure cookers, can continue their work for weeks or months after an initial exposure, refining antibodies further. This is one reason the timing between vaccine doses can matter: the immune system may be using the interval to improve the quality of the response, not merely to rest.

A key player in this prolonged refinement is the follicular dendritic cell (FDC), which is neither a classical dendritic cell nor a typical immune cell. FDCs sit in follicles of lymphoid tissues and act as display boards for antigen. They can retain antigen in the form of immune complexes for long periods, on the scale of months, holding onto molecular evidence of an encounter long after the pathogen itself has been cleared from the bloodstream. The antigen is not necessarily floating around freely; it is pinned, preserved, and rationed out to B cells that compete for it. In effect, the immune system keeps a sample of the problem around long enough to keep improving its solution.

Even more intriguing, FDCs do not just hold antigen; they can cycle immune complexes in ways that may protect antigen from degradation while keeping it available. The details are technical, but the intuition is simple: if you want to keep a fragile object on display for months, you do not leave it in the sun and rain. You store it and bring it out as needed. By controlling antigen availability, FDCs can extend the period during which germinal centers can select better B cell variants.

The COVID-19 pandemic provided a rare, intense window into these long timelines in humans. Immunology often relies on animal models because lymph nodes are not easily sampled repeatedly in healthy people. During the rollout of SARS-CoV-2 mRNA vaccines, a number of research groups followed immune responses in extraordinary detail, including direct sampling of lymph nodes in volunteers. These studies showed that antigen-specific germinal center responses could persist for months after vaccination, and that B cell lineages continued to accumulate somatic hypermutations over similar time scales. In other words, the immune system kept iterating on antibody design long after the soreness in the arm had disappeared.

This matters for a reason that is both technical and deeply practical: antibodies can improve not only in affinity but in *breadth*. A virus such as SARS-CoV-2 changes over time, accumulating mutations in the very proteins antibodies tend to target. An antibody response that is narrowly focused may be powerful against the original strain but brittle against variants. Continued germinal center activity and continued selection can sometimes yield antibodies that recognize conserved features of the virus, providing better cross-protection. That does not mean every response becomes broadly protective; biology rarely offers guarantees. But it does mean that time itself can be an ally, allowing the immune system to explore more of the "solution space" of possible antibodies.

This long refining period also helps explain why the immune system sometimes seems inconsistent in the duration of protection. The durability of immunity is not solely a question of whether memory cells exist. It depends on how many long-lived plasma cells were established, how strong their bone marrow niches are, what level of antibody is maintained, and how rapidly that antibody decays. It depends on whether memory cells receive enough homeostatic signals to survive. It depends on the pathogen's behavior and the site of infection. And it depends on how much antigen persisted long enough to drive extended selection and maturation.

These are many moving parts, but a simple quantitative picture brings them into focus. Imagine measuring a person's antibody level against a virus over time. After infection or vaccination, the level rises rapidly during the expansion phase, then falls during contraction as short-lived plasma cells die, and then settles into a plateau maintained by long-lived plasma cells. That plateau can be high and flat, or low and slowly declining. The difference can mean the difference between sterilizing immunity (blocking infection outright) and protection mainly against severe disease. It can also determine whether a booster is needed.

A chart can capture these phases without pretending to predict exact numbers:

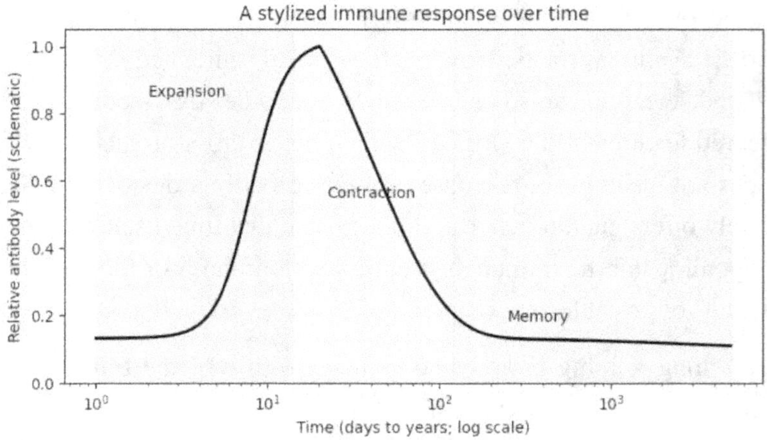

The important point in such a graph is not the particular curve; it is that the immune system has at least two kinds of persistence. One is the persistence of cells that can respond again (memory B cells and memory T cells). The other is the persistence of molecules already doing the job (antibodies maintained by long-lived plasma cells). If antibody levels stay above a protective threshold, a reinfection may be blocked quickly. If antibody levels fall below that threshold, infection may occur but be contained more efficiently because memory cells respond faster than naive cells.

A historical anecdote shows how long-lasting immunity can be, and how it can be measured. Smallpox vaccination, using related viruses such as vaccinia, was famously effective and ultimately enabled the eradication of smallpox. Long after routine smallpox vaccination stopped, immunologists were still able to find evidence of robust immune memory in vaccinated individuals decades later. Studies in the early 2000s examined people vaccinated many years before and found persistent T cell responses and antibody levels that declined slowly, not precipitously. The details of smallpox immunity are complex and cannot be copied wholesale to every vaccine, but the broad lesson remains: immune memory can be extraordinarily durable when the right kinds of long-lived cells and niches are established.

The same persistence that protects can also haunt. In allergies, long-lived plasma cells can maintain IgE or IgG antibodies that keep the body hypersensitive to harmless substances. In certain autoimmune diseases, long-lived plasma cells can sustain antibodies against the body's own components, contributing to chronic damage. The "long wait" is not morally good; it is a strategy. Once established, it can be hard to undo because it is literally built into the body's cellular architecture.

The immune system's patience also raises a philosophical question that is more than wordplay: what does it mean for a cell to "remember"? A memory T cell is not constantly replaying an antigen. A memory B cell is not constantly sampling the world for its specific target. Memory is stored in receptor specificity, in epigenetic readiness, in survival programs, in tissue positioning, and in the niches that keep certain cells alive. It is memory as a set of biases: biases in who survives, where they live, how quickly they respond, and what they are capable of becoming.

There is a quiet elegance to this arrangement. It accepts that perfect vigilance is impossible. It accepts that the world changes and pathogens change. Instead of trying to maintain a static, exhaustive defense, it maintains a set of capabilities that can be reactivated and adapted. Some of those capabilities are immediate, like circulating antibody. Some are local, like resident cells in barrier tissues. Some are latent, like central memory cells ready to proliferate. The waiting is active in the way a well-kept instrument waits: not making sound, but staying in tune.

A person rarely feels this tuning. Most of the time, it is only visible in its absence: recurrent infections, unusually severe disease, vaccines that fail to take, immune systems depleted by chemotherapy or immune disorders. But when it works, the body carries an archive of molecular solutions, each one maintained not by a single mechanism but by overlapping forms of cellular and anatomical commitment. Somewhere in bone marrow niches, plasma cells continue their slow, steady work. Somewhere in tissues, resident memory cells keep their posts. Somewhere in lymph nodes, memory cells circulate through familiar corridors, quietly surviving on small signals, ready to become loud.

Chapter 5

The Mathematics of Survival

To an observer, the immune system appears to possess intent, hunting pathogens with what looks like tactical precision. Yet, this behavior emerges not from a central brain, but from a decentralized calculus where biochemical thresholds and probability distributions distinguish harmless noise from existential threats. By treating cellular interactions as information processing, we reveal the rigid mathematical logic that governs whether the body chooses to tolerate a stimulus or launch a destructive attack.

5.1 The Calculus of Activation

A cell lives in a world of bumps.

At any instant, thousands of molecules strike its surface, drift away, return, collide again. Many of these encounters are meaningless. Some are actively misleading. A healthy body is full of perfectly ordinary proteins that, from the viewpoint of chemistry, can look uncomfortably similar to those made by microbes. If immune cells reacted to every brief contact, the result would not be vigilance but panic: chronic inflammation, tissue damage, and the slow self-sabotage of autoimmunity.

So an immune cell faces a survival problem that resembles a familiar one in everyday life: how to treat a stream of interruptions. A single phone buzz might be a pocket mistake. Ten pings in ten seconds could be a crisis. The important detail is not only *what* happened, but *how long it persisted* and *how consistently it repeated.* In the immune system, the equivalent of "time on task" is the duration and stability of a molecular bond between a receptor on the immune cell and a ligand (a binding partner) on another cell.

For T cells, the crucial receptor is the T cell receptor (TCR). The TCR does not grab whole pathogens directly. It reads short peptide fragments presented by other cells on molecular display stands called MHC (major histocompatibility complex). A T cell's job is to decide whether a presented peptide is a harmless self fragment, a fragment from a past infection that no longer matters, or a sign of an active threat. This decision often must be made quickly: in minutes to hours, not days. Yet the discrimination must also be exquisitely specific. The typical T cell must ignore an enormous background of self peptides while still being able to respond when only a few threatening peptide—MHC complexes are present.

It is tempting to imagine the TCR as a switch: bind the right thing, flip to "on." But many experiments have made that picture hard to sustain. The binding differences between "right" and "almost right" ligands can be modest. Sometimes the affinity (how tightly two molecules bind) is not dramatically different. What often differs more consistently is *dwell time*: how long, on average, the receptor stays bound before the ligand slips away. A useful way to think about a T cell, then, is not as a simple switch but as a device that performs a kind of calculus: it accumulates evidence over time, discounts fleeting contacts, and commits only after the signal has survived repeated opportunities to fail.

That sounds abstract until one remembers the physics of the cell membrane. The surfaces of cells are not rigid plates; they are crowded, jostling seas of proteins anchored in a fluid lipid layer. Two cells meeting in a lymph node touch, detach, and touch again as they migrate. Even during a "stable" contact, receptor—ligand pairs are constantly breaking and reforming. In such a setting, a decision rule based on instantaneous binding would be almost indistinguishable from noise. The only way to create reliability from that turbulence is to require persistence.

A classic idea capturing this requirement is *kinetic proofreading*, introduced into T cell signaling models most influentially by Thomas McKeithan (1995). The phrase comes from molecular biology's broader preoccupation with error correction. Proofreading, in general, means adding steps that cost time and energy so that incorrect matches are likely to fall apart before they can trigger an irreversible outcome. In DNA replication, proofreading enzymes do this by repeatedly checking base pairing. In T cell signaling, the "checks" are biochemical steps that must occur while the receptor remains bound. If the ligand dissociates too quickly, the chain of steps collapses, and the cell resets.

One can picture the process like a nightclub with a series of doors. Entering the building is easy: many people can walk in. But to reach the main room, you must pass through several checkpoints. If you leave the line at any point, you go back outside. A person who lingers and keeps moving forward eventually gets inside; someone who merely bumps the first door and backs away never reaches the party. The point is not cruelty; it is selectivity. The immune system, like any security system, gains specificity by refusing to be impressed by momentary contact.

Mathematically, kinetic proofreading turns small differences in dwell time into large differences in downstream signaling. Suppose each checkpoint step occurs with some rate, and the ligand can fall off with another rate. If the bond breaks quickly, the probability of surviving many steps is tiny. If the bond lasts longer, the probability rises steeply. The same principle holds even when the underlying chemistry is messy. The cell does not need to compute probabilities explicitly. It only needs a network of reactions arranged so that progress depends on sustained binding. The "calculus" is embedded in the kinetics.

This time-based strategy helps resolve a famous tension in T cell biology: the need to be both sensitive and specific. Sensitivity demands that a T cell respond even when only a few agonist (truly activating) ligands are present. Specificity demands that it *not* respond to the vast number of self ligands. If activation depended mainly on quantity, then a large pile of weak self ligands could overwhelm the system. Proofreading shifts the emphasis toward quality: a rare ligand that stays bound long enough can drive progress through checkpoints, while abundant short-lived ligands repeatedly fail.

The story becomes more vivid when one adds a second observation: inside individual T cells, many downstream responses look *digital.* Molecules such as ERK (a kinase in the MAPK pathway) and transcription factors such as NFAT can switch between low and high states in an all-or-none way at the single-cell level, even while the population as a whole shows graded behavior. Géraldine Altan-Bonnet and Ronald Germain (2005) helped cement this view by modeling how feedback control can produce robust, switch-like ERK activation while preserving discrimination. A digital decision is useful because it prevents half-measures. A partially activated killer T cell is not a safe compromise; it is a hazard. Either the cell commits to a program or it remains quiet.

Digital behavior, however, creates its own problem. If the internal decision is all-or-none, how can an immune response be tuned to match the size of a threat? One answer is that the tunable part moves from amplitude to *frequency*: not how strongly each responding cell responds, but how many cells respond and how often they do so. A population can produce a smooth, graded immune reaction simply by varying the fraction of cells that cross the threshold. That is a statistical solution to a biochemical problem: a million binary decisions can approximate a dimmer switch.

These ideas, though elegant, can sound like they belong on a whiteboard rather than in a lymph node. A concrete scene helps. Consider what happens during the early stages of an infection, when a dendritic cell (a professional antigen-presenting cell) arrives in a lymph node carrying molecular evidence of a pathogen. Naïve T cells patrol the node, making brief contacts with dendritic cells as they move. Most encounters are quick and unproductive. Yet when a T cell's receptor recognizes a relevant peptide—MHC complex, the contact can stabilize. The T cell slows, spreads, and begins forming a specialized interface sometimes called an immunological synapse. In that interface, receptors cluster and signaling molecules reorganize. It is not merely a stronger handshake; it is a restructuring of the conversation.

Now imagine the difference between two ligands. Ligand A binds the TCR but falls off in a second or less. Ligand B stays bound for several seconds. Those numbers seem trivial in human time. In biochemical time, they are enormous. A multi-step phosphorylation cascade can proceed through several modifications in a few seconds if conditions are right. If the ligand falls off too soon, the cascade stalls and reset mechanisms (including phosphatases that remove phosphate groups) erase progress. If the ligand persists, the cascade reaches a point where positive feedback accelerates the remaining steps, pushing the cell over a threshold into a digital "on" state. The T cell has effectively measured time without owning a clock. The clock is the lifetime of the bond.

Kinetic proofreading is not the only way to implement this measurement. More recent work has emphasized spatial organization: the cell's surface is not uniform, and the balance of kinases and phosphatases can change dramatically within tiny close-contact zones where membranes are pressed together. In these zones, bulky phosphatases may be physically excluded, tilting the chemistry toward activation. Dwell time matters here too, because a short-lived interaction may not persist long enough within a favorable microdomain to accumulate modifications. Whether one emphasizes sequential biochemical steps or spatially regulated chemistry, the same computational theme remains: discriminate by requiring persistence.

This returns us to the question of noise. In an electronic circuit, noise is unwanted variability in voltage. In immunology, noise has several faces. There is physical noise: random collisions and fluctuations in membrane organization. There is molecular noise: low copy numbers of key signaling proteins, causing variability from cell to cell. There is informational noise: peptides that resemble danger but are actually self. A T cell's problem is to avoid being fooled by all three at once.

The key trick is to convert a noisy stream of brief inputs into a smoother internal variable. Engineers call this *integration.* A leaky bucket is a classic metaphor. Imagine a bucket under a dripping faucet. Each drip adds a bit of water, but there is a small hole at the bottom. If drips are sporadic, the bucket never fills; water leaks away between drips. If drips come fast enough, the water level rises and eventually reaches a marked line. That line is a threshold: above it, an alarm goes off. Below it, nothing happens. The hole is crucial. Without it, any drip, no matter how rare, would accumulate forever, and eventually everything would trigger. The immune system cannot afford such perfect memory at the level of activation, because it would be haunted by every incidental encounter.

Inside T cells, phosphatases play the role of the leak, continuously undoing phosphorylation events unless kinases keep renewing them. Feedback loops can tighten or relax the leak dynamically. A strong sustained signal can temporarily reduce leakiness or amplify filling, creating a rapid threshold crossing. A weak or transient signal cannot keep up with the loss. The resulting behavior is not mystical. It is the everyday logic of differential equations written in protein.

A historical anecdote makes the stakes tangible. In the early days of organ transplantation, surgeons and immunologists confronted a brutal fact: the immune system can treat a life-saving graft as a lethal invader. The intensity of rejection depends not only on the presence of foreign antigens but on the conditions under which immune cells encounter them. A transplanted organ arrives with inflammation, tissue damage, and activated antigen-presenting cells. Those contextual cues can provide sustained, reinforcing signals that push lymphocytes past activation thresholds. By contrast, exposure to foreign antigens in a non-inflammatory setting can sometimes lead to tolerance rather than attack. The difference is partly collective and partly contextual, but it also reflects the same cellular arithmetic: sustained, multi-signal engagement drives commitment; fleeting, unreinforced encounters tend to be ignored or even used as lessons in restraint.

Kinetic proofreading offers an especially satisfying explanation for why "almost dangerous" ligands can be actively disruptive. If a ligand binds for an intermediate time, it might initiate early steps but fail to complete them, tying up receptors and signaling machinery without delivering a full activation. François and Altan-Bonnet (2016) emphasized that strict discrimination can generate *antagonism*: near-threshold ligands that reduce the response to true agonists. In human terms, it is the person who keeps ringing your doorbell and walking away. Each ring is not enough to make you open the door, but it prevents you from hearing the knock that matters. Such antagonism is not a design flaw so much as a trade-off that appears when a system tries to be simultaneously fast, sensitive, and specific.

There is also an emotional aspect to this story, because the immune system's thresholds are not fixed. They are tuned by history, metabolism, and context. A T cell recently exposed to certain cytokines can lower its threshold; another under chronic stimulation might raise it or become exhausted, requiring stronger signals to respond. Even within one person, thresholds differ across tissues: the gut, constantly exposed to food antigens and microbiota, cannot run the same trigger settings as the bloodstream. The "calculus" is therefore adaptive. It is closer to an adjustable thermostat than a tripwire.

Yet even an adjustable thermostat relies on the same core principle: do not react to every gust of cold air. Wait for a sustained change. In the immune system, waiting is not passivity. It is active computation performed by molecules that spend their lives in motion. The cell does not ask, "Is this molecule foreign?" in a single glance. It asks, "Does this interaction endure long enough to justify escalation?" It asks, "Is the evidence persistent enough to survive repeated attempts to erase it?" It asks, "Does the signal arrive with the right timing and reinforcement to merit commitment?"

The logic of this becomes easier to see when one draws the difference between two time traces: one representing noise, the other representing a genuine signal. Noise produces frequent spikes: contacts happen, small bursts of phosphorylation occur, but they decay. A genuine threat produces sustained engagement: accumulation rises and crosses a threshold. The cell's decision machinery is built to respond to the second pattern and ignore the first.

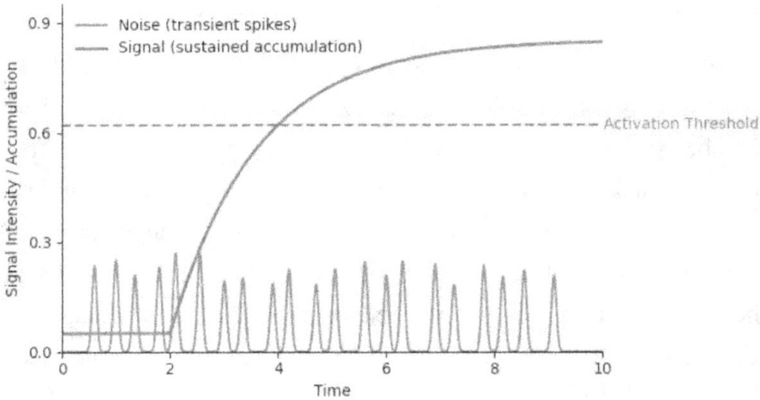

If one had to name the quiet hero of immune decision-making, it might be the humble reset. Cells are full of enzymes whose job is to erase modifications, dismantle transient complexes, and return the system to baseline. These erasers are sometimes portrayed as brakes. They are better understood as accountants. They prevent the cell from being tricked by partial evidence. They ensure that activation requires not merely contact, but *continued contact in the face of relentless undoing*.

This is why a T cell can be simultaneously impatient and cautious. It is impatient because once the evidence persists, it can cross a threshold quickly, recruiting feedback loops that make the response decisive. It is cautious because the evidence must endure long enough to outpace forgetting. In that balance lies a practical form of wisdom: a disciplined refusal to confuse passing noise with a reason to burn the house down.

5.2 Logic Gates and the Chemical Internet

A single immune cell can make a careful decision and still be disastrously wrong.

That is not a criticism of the cell; it is a statement about the kind of world immune systems inhabit. A T cell might correctly detect a viral peptide on one dendritic cell, yet the infection could already be contained, the tissue could be calm, and a full-scale inflammatory response would do more harm than good. Or a cell might miss a threat simply because the relevant encounter was rare. Life at the microscopic scale is ruled by sampling error: small numbers, brief meetings, incomplete information.

The body's answer is not to demand perfection from each cell but to turn immunity into a collective computation. Immune cells do not merely *act*; they *broadcast*. They release and sense cytokines, small signaling proteins that diffuse through tissues and lymph nodes. Cytokines are the immune system's short messages: "danger here," "help needed," "slow down," "divide," "differentiate," "stop." No single message is trusted on its own. What matters is the pattern, the timing, the local concentration, and the identity of the senders and receivers.

Calling this a "chemical internet" risks sounding cute, but the analogy is practical. The internet is not one computer thinking very hard; it is many devices sending packets, repeating signals, amplifying some, suppressing others, and making global outcomes from local rules. A lymph node behaves in the same spirit. Millions of cells, each with partial information, exchange cytokines that act like packets with different priorities and ranges. Some are whispers that fade within a few cell diameters; others are loud, systemic alerts that reach bone marrow and brain. There is no central controller. Yet coordinated behavior emerges: fever, antibody production, killer-cell expansion, tissue repair, and, crucially, restraint.

Restraint is the underappreciated achievement. The immune system's most dangerous tools—killing infected cells, releasing toxic granules, recruiting swarms of inflammatory cells—must be deployed only when necessary. Every day, the body contains countless potential triggers: dying cells, benign microbes, food antigens, and self proteins modified by normal metabolism. A system that acted on single cues would constantly ignite. To avoid that, immune cells use something that looks very much like logic.

Logic, in the everyday sense, means rules for combining evidence. In electronics, logic is embodied in gates: an AND gate produces an output only when *both* inputs are present; an OR gate when *either* is present; a NOT gate inverts a signal. Cells do not carry circuit diagrams, but their receptors and signaling pathways can be arranged so that one signal is insufficient. In T cells, the best-known example is co-stimulation: the T cell receptor may detect a peptide, but full activation usually requires additional signals from molecules such as CD28 engaging ligands on antigen-presenting cells. Without that second input, a T cell may remain inactive or even become tolerant.

Cytokines extend this logic from a one-on-one conversation to a community decision. They create distributed AND gates: "antigen recognition *and* inflammatory context," "local damage *and* interferon," "enough activated neighbors *and* enough growth factor." The immune response, on this view, is not a single trigger but a vote tallied across many cells, where cytokines are ballots counted in real time.

A vivid example comes from a molecule that immunologists have called, without exaggeration, a growth factor, a hormone, and a social signal: interleukin-2 (IL-2). IL-2 is produced primarily by activated T cells and strongly promotes T cell proliferation and survival. It is also consumed aggressively by cells that express the high-affinity IL-2 receptor, including regulatory T cells (Tregs), which specialize in suppression. IL-2 therefore does not behave like a simple broadcast that everyone hears equally. It behaves like a message in a crowded room where some listeners have megaphones and others have earmuffs, and where the loudest listeners can also *absorb* the sound.

The result is subtle but consequential: IL-2 can act as a quorum-like signal. A single T cell producing a little IL-2 may not raise the local concentration enough to sustain its own expansion, especially if nearby cells soak it up. But if many T cells become activated at once, IL-2 levels can rise sharply, tipping the local environment toward proliferation. In effect, the population asks: are there enough of us seeing something worth responding to? The answer is not spoken; it is measured as a concentration.

This sort of collective threshold helps explain a long-standing observation in immunology that is easy to miss because it is now taken for granted: the immune system often prefers *coherent* danger. One infected cell in a tissue might be handled quietly by local defenses. A spreading infection, with many infected cells and damaged tissue releasing alarm signals, drives a different mode entirely. Cytokines are the medium through which coherence is detected.

Historical anecdotes can make this less abstract. In the late twentieth century, clinical medicine repeatedly encountered a terrifying syndrome in which the immune system's communication network runs out of control: sepsis and septic shock. Although the details are complex and still debated, one broad lesson emerged from decades of intensive care research: it is not only the pathogen that kills. The host response can become a runaway cascade of cytokines, sometimes described in popular writing as a "cytokine storm." This phrase can be overused, but the underlying point is real enough to haunt clinicians: cytokines can synchronize immune behavior so effectively that, if the synchrony is misguided or unchecked, the result is systemic collapse—fever, low blood pressure, organ failure.

That is the dark side of a chemical internet: networks can amplify. The bright side is that networks can also impose safety checks. In well-functioning immunity, cytokine circuits often include built-in brakes. A signal that turns something on also triggers inhibitors that turn it off later. This pattern is familiar to anyone who has watched a thermostat: heating raises the temperature, and the rising temperature turns the heater off. In cytokine signaling, inhibitory proteins such as SOCS (suppressors of cytokine signaling) can be induced by the very pathways they inhibit, creating a timed self-limitation. Géraldine Altan-Bonnet and colleagues have argued persuasively that cytokine communication is best understood quantitatively, with attention to receptor numbers, diffusion, consumption, and feedback inhibition. In that quantitative view, safety is not a moral virtue; it is a dynamical property of a network.

A simple, human-scale analogy helps. Imagine a neighborhood deciding whether to evacuate because of a potential gas leak. One person smelling something odd is not enough. People check with neighbors. Some are more sensitive to smells; some are more cautious. If many neighbors report the same thing and the smell persists, the group evacuates. If the reports are scattered and transient, people stay. Now add a key complication: some people have a role in preventing panic, perhaps a trained safety officer who insists on confirmation and shuts down rumors. In the immune system, cytokines carry the "smell reports," and regulatory cells act like those officers, damping overreaction by consuming shared growth factors and secreting inhibitory cytokines such as IL-10 or TGF-β.

The "internet" metaphor becomes especially powerful when one considers that cytokines do not merely report information; they *shape* the network of who talks to whom. A cell can change its receptor expression, effectively tuning in or tuning out. A T cell exposed to one cytokine milieu becomes a different kind of T cell, producing a different set of cytokines in return. The network rewires itself as it runs.

Take the differentiation of helper T cells. A naïve CD4 T cell encountering antigen can develop into subsets often labeled by their characteristic cytokines and functions: Th1 cells that produce interferon-γ and support responses to intracellular pathogens; Th2 cells producing IL-4, IL-5, and IL-13 associated with responses to parasites and allergies; Th17 cells producing IL-17 involved in responses to extracellular bacteria and fungi and implicated in inflammatory diseases; T follicular helper cells that support antibody-producing B cells. These categories are simplifications of a more continuous landscape, but the key point holds: cytokines guide fate decisions. In other words, the messages are not only about "act now" but about "become this kind of actor."

This helps resolve a puzzle that otherwise seems paradoxical: how the immune system can reuse the same cells for different kinds of danger without confusing itself. The same naïve T cell pool can produce very different outcomes depending on cytokine context. Context is not an afterthought; it is part of the computation. Antigen recognition tells a cell *what* is being seen. Cytokines tell it *what sort of world* it is in—a world of viruses, worms, damaged epithelium, or sterile injury. The final behavior emerges from combining these signals in something like an AND gate.

There is another, less intuitive part of the story: communication is limited by physics. Cytokines diffuse, but they are also consumed, degraded, and trapped by receptors. A cytokine message does not fill the body evenly like dye in a bathtub. It forms gradients. Near the secreting cell, concentration can be high; a few cell diameters away, it may drop sharply. This makes cytokine communication partly local, partly regional, and only sometimes systemic. The immune system uses that fact. It can make local decisions without alarming the whole organism, and it can escalate to systemic signaling only when local containment fails.

Diffusion and consumption also mean that cytokine signals are competitive. Cells do not just listen; they compete for the message. High receptor expression is a way to capture a larger share of a limiting cytokine, much as a faster internet connection can capture more bandwidth. This turns cytokine signaling into a resource economy. In the case of IL-2, cells expressing high-affinity receptors can deprive others, thereby shaping which clones expand. Regulatory T cells, with their constitutively high IL-2 receptor expression, are especially well placed to modulate this economy, acting as sinks that prevent small, potentially autoreactive responses from snowballing.

Information theory offers a language for these limitations that is both sobering and clarifying. Any communication channel has a capacity: a limit to how much information can be transmitted reliably in the presence of noise. In cytokine signaling, noise arises because cells differ in receptor numbers, signaling protein abundance, metabolic state, and microenvironment. Even genetically identical cells can respond differently. Experiments on pathways such as IL-6 to STAT3 have shown that cell-to-cell variability can constrain how precisely a cell can infer the concentration of a cytokine from its internal response. Yet biology does not necessarily need high precision. A smoke alarm does not need to measure particulate concentration to three decimal places; it needs a robust threshold. Often, the immune system seems built for decisions that are coarse but safe: respond or do not respond; proliferate or do not; differentiate in one direction or another.

Robustness, however, can conflict with information. A pathway that aggressively saturates and thresholds may be resistant to fluctuations, but it may also discard subtle distinctions. That trade-off is familiar in engineering: hard thresholds reject noise but also compress detail. In immune communication, this trade-off may be advantageous. If the cost of a false positive is tissue damage and the cost of a false negative is infection, the "best" threshold depends on context, past experience, and even life stage. The immune system appears to adjust these thresholds through feedback inhibition, receptor regulation, and cross-talk between pathways. Instead of relying on a single, fragile readout, it uses ensembles of cues.

Temporal patterns add another layer of logic. Some cytokine-induced transcription factors move in pulses, not smooth ramps. NF-κB, a central regulator of inflammation, is famous for oscillatory nuclear translocation in response to certain stimuli. Pulses can encode information in timing: how often the nucleus is visited, how long it stays, whether it returns quickly after a previous visit. In social terms, it matters whether someone calls once and hangs up or calls repeatedly over an hour. Timing can differentiate persistent danger from a transient blip.

Yet time coding also faces constraints. A system that uses pulses must reset between pulses; otherwise signals blur. Experimental work with pulsatile stimulation has shown that too-frequent inputs can reduce effective signaling because negative feedback and refractory dynamics prevent full reset. That is a deep point about biology's computing limits: cells cannot respond arbitrarily fast simply because chemistry is fast. They are limited by the time constants of their own safety circuits.

All of this might still feel distant from lived experience until one considers the daily negotiations between the immune system and the microbiome. The gut is packed with bacteria that are not merely tolerated but essential. These microbes produce molecules that could, in other contexts, look alarming. The gut immune system must distinguish between "normal microbial presence" and "breach of the barrier." Cytokines are central to this distinction. When the epithelial barrier is intact, signals promoting tolerance and containment dominate. When the barrier is damaged, epithelial cells and innate immune cells release cytokines and chemokines that recruit inflammatory cells and promote more aggressive responses. The same bacterial molecules can be interpreted differently depending on the cytokine context. The decision is distributed: epithelial cells report damage, innate cells report pattern recognition, T cells report antigen specificity, and the cytokine environment tallies the combined evidence.

The logic-gate metaphor earns its keep here. A gut T cell does not launch an inflammatory program merely because it sees a bacterial antigen. It also needs confirmation that the antigen is appearing in the wrong place, at the wrong time, with the wrong accompanying signals. That is an AND gate implemented by biology: antigen *and* danger context. When these gates fail, inflammatory bowel disease can result, a costly illustration that "computation errors" in immunity are not abstract.

There is a final twist. In electronics, gates are designed; in immunity, they are evolved and learned. Learned, in the limited but real sense that immune cells alter their responsiveness based on experience. Cytokine exposure can tune receptor expression and downstream sensitivity. Memory T cells, shaped by past infections and vaccinations, often require less co-stimulation and can respond more rapidly. That is a rational adjustment: once a pathogen has been confirmed as dangerous, the system lowers the bar for recognizing it again. But lowering the bar too far would be dangerous, so tuning is rarely global. It is specific, contextual, and often constrained to particular tissues where the memory was formed.

The immune system, then, does not simply identify enemies. It weighs evidence in a noisy world, distributes risk across many agents, and uses chemical messages to coordinate action while enforcing checks. Cytokines are not mere "signals" in the casual sense. They are the medium through which countless small decisions become a coherent response, through which a local skirmish can stay local or escalate to an organism-wide mobilization. In a healthy response, the chemistry feels almost humane: urgency without hysteria, coordination without central command, and the continual willingness to stand down when the votes do not add up.

5.3 The Strategic Value of Noise

Engineers tend to treat noise as an insult. A radio crackles; a microphone hisses; a photograph becomes grainy in low light. The instinct is to suppress variation, to tighten tolerances, to make every component behave identically. Biology did not get that memo. Living systems are noisy by construction: molecules arrive late, reactions misfire, genes turn on in bursts, cells differ from their neighbors for reasons no microscope can fully enumerate. Immune systems are especially noisy because they operate at the edge of detection, trying to pick out rare, dangerous patterns in a sea of innocuous ones.

At first glance, that sounds like a recipe for disaster. If immune cells fluctuate, surely they will sometimes activate when they should not, or fail when they should. Yet the immune system has survived long enough to be written about precisely because it has found ways to turn randomness into resilience. Noise is not only a limitation; it can be a strategy.

There are two different senses in which noise can help. The first is statistical and almost philosophical: in an unpredictable world, uniformity is a weakness. The second is mechanistic: in systems built around thresholds and feedback, a little randomness can improve detection and coordination, much as a small amount of jostling can help a stuck drawer slide open. These senses are related but not identical, and immunity uses both.

A helpful starting point is to notice that the immune system's most consequential decisions are often binary at the level of an individual cell. A T cell either commits to a program of proliferation and cytokine secretion, or it does not. A B cell either enters a germinal center to refine antibodies, or it remains quiescent. A macrophage either turns on an inflammatory state or stays in a repairing, homeostatic mode. The internal circuitry that produces these binary outcomes typically involves thresholds, positive feedback, and negative feedback, the biochemical equivalents of a snap-action switch.

A snap-action switch is satisfying because it prevents half-measures, but it is also vulnerable to fluctuations near the threshold. If the signal is right at the decision boundary, small variations in molecule numbers, reaction rates, or timing can determine whether the cell flips. In an engineered device, that would be an annoyance. In an immune population, that variability can be a form of insurance.

Consider a scene in which a new pathogen enters the body. The host has never encountered it before, so there is no pre-existing memory specialized for its antigens. The immune system must improvise using a diverse pool of naïve lymphocytes, each with a different receptor. The first cells to recognize the pathogen may do so weakly. The context may be ambiguous: early in infection, danger signals may be modest, and antigen may be scarce. A rigid, uniform threshold could lead to a uniform failure: nobody activates until it is too late. A noisy population, by contrast, contains outliers. Some cells, by chance, have slightly higher receptor expression, slightly more favorable signaling protein levels, slightly lower inhibitory tone. Those cells may cross the threshold early. They may buy time for the rest of the system, recruiting help through cytokines and reshaping the environment so that others can follow.

This is one meaning of *bet-hedging*, a concept borrowed from evolutionary biology. A bet-hedging organism, or system, deliberately produces variability in its offspring or components so that at least some will survive unforeseen challenges. Desert plants that germinate their seeds across multiple years hedge against unpredictable rainfall. Bacteria that produce a small fraction of dormant "persister" cells hedge against antibiotics. The immune system hedges by maintaining diversity across receptor sequences, of course, but also by allowing variability in activation thresholds and response programs.

Bet-hedging is easiest to grasp when the danger is rare but catastrophic. Imagine a pathogen that can only be defeated if the immune response is mounted within a narrow time window. A uniform response tuned to avoid false alarms might miss that window. A noisy response would occasionally trigger prematurely, which is costly, but it increases the probability that someone fires in time when the threat is real. Natural selection, indifferent to comfort and attentive only to survival, will often favor strategies that look wasteful from an engineering standpoint but are rational under uncertainty.

A concrete medical example is the family of viruses that can establish rapid, systemic infections with high mortality in immunologically naïve hosts. Historical accounts of outbreaks often describe how quickly a patient can deteriorate: fever, malaise, then sudden worsening. In such settings, the immune system benefits from having some cells that are willing to respond to weaker cues, even if those cues are sometimes mistaken. The alternative is a system that is beautifully calm and precisely regulated right up until the moment it is overwhelmed.

The immune system does not leave this hedging to chance alone. It embeds it in the way immune cells develop and circulate. Even within a population of genetically identical T cells, there is measurable variation in the abundance of key signaling molecules. Some of that variation is stable for hours or days; some fluctuates. Cells also occupy different microenvironments, encountering different cytokine backgrounds, nutrient availability, and mechanical constraints. That means the same antigenic stimulus can produce a spread of responses: some cells remain silent, some activate fully, some activate transiently and then stop.

This is not merely tolerated; it can be used. Population-level outcomes often depend on the *fraction* of cells responding rather than the intensity of any one cell. A thresholded system can translate noisy inputs into a tunable response by modulating response probability. Instead of asking each cell to be a perfectly calibrated instrument, the immune system treats each cell as a vote that can be swayed by context. The immune response becomes a statistical object.

Information theory provides a rigorous way to talk about this. In a noisy channel, one cannot reliably transmit an arbitrary number of distinguishable messages. The immune system, facing noise in receptor binding and intracellular signaling, may not be able to encode fine-grained information about antigen quantity in a single molecular readout. But it can often reliably convey a coarser message: "below threshold" versus "above threshold." If more detail is needed, it can be distributed across time (pulses) and across cells (fractions responding). The message is not "the antigen concentration is 17.3" but "this is real enough to act on, and the community agrees."

The mechanistic side of noise becomes especially interesting when one looks at signaling dynamics. Many immune pathways are not smooth. They pulse. NF-κB can enter and leave the nucleus in oscillations. Calcium signals can spike. These temporal patterns can carry information, but they also arise partly because feedback loops impose rhythms. Negative feedback produces overshoot and return; delayed inhibition produces oscillation. When stimuli arrive in pulses, the system must reset between them. If pulses are too frequent, reset fails, and responses can diminish rather than increase. That failure is not a bug; it is a safety feature. It prevents a cell from interpreting incessant, jittery stimulation as a steadily increasing threat. It forces the system to care about temporal structure.

Noise interacts with such dynamics in nontrivial ways. Near a threshold, a small fluctuation can determine whether a pulse crosses into a full response. In some nonlinear systems, the right amount of noise can even *enhance* detection of weak periodic signals, a phenomenon known as stochastic resonance. The basic idea is counterintuitive but simple: if a signal is too weak to cross a threshold on its own, and noise is too weak to matter, the system never triggers. If noise is enormous, triggers happen randomly and bury the signal. But at an intermediate noise level, the weak signal biases the timing of threshold crossings, making the output more synchronized with the signal. Some immunologists have speculated that analogous effects could occur in immune recognition microdomains, where clusters of receptors sample a fluctuating background of self ligands. The evidence for this in immunity is less established than the core principles of kinetic proofreading and feedback control, but the conceptual lesson is valuable: thresholds do not only suppress noise; they can also harness it.

Noise is also tied to another kind of robustness: robustness against deception. Pathogens are not passive. They evolve ways to interfere with immune signaling, to suppress cytokines, to mimic host molecules, to decoy receptors, to hide antigens. A perfectly uniform immune response can be an easy target. If a virus evolves a protein that blocks a particular signaling step, and every relevant immune cell relies on that step in the same way, the virus has found a master key.

Variation makes master keys less effective. If different cells rely on slightly different parameter regimes, or activate through partially redundant pathways, then a pathogen's interference will be incomplete. Some cells will still respond. The immune system has many forms of redundancy, but noise creates a subtler kind: not redundancy of components, but redundancy of *states*. A pathogen may block the average pathway, but it may fail to block the outliers that matter most during crisis.

A historical anecdote illustrates the practical consequences of outlier cells. When smallpox vaccination was first widely deployed in the nineteenth century, physicians observed something that modern immunology now treats as obvious: not everyone responded identically. Some developed strong "takes" at the vaccination site and robust protection; others had mild reactions and weaker immunity; a few had adverse effects. At the time, this variability was frustrating and poorly understood. Today it is seen as an unavoidable feature of immune systems shaped by genetics, age, nutrition, prior infections, and stochastic cellular processes. Variability is not merely a nuisance in public health; it is a property that can protect populations. If a pathogen spreads through a community, heterogeneous immune responses mean the pathogen encounters a patchwork of environments. That patchwork can slow transmission and reduce the chance that a single viral variant dominates. On the other hand, the same heterogeneity can leave vulnerable individuals at higher risk, which is why vaccines and therapies aim to reduce dangerous variability while preserving beneficial diversity.

At the cellular level, the tension between helpful and harmful noise is especially stark in autoimmunity. A noisy threshold system means that, occasionally, a self-reactive cell might cross the activation boundary when it should not. The immune system reduces this risk through multiple layers: central tolerance during lymphocyte development, peripheral tolerance mechanisms, regulatory cells, and co-stimulation requirements. These layers can be seen as ways of shaping the noise: allowing beneficial variability in sensitivity while preventing rare, catastrophic misfires. The system tries to keep the distribution broad enough to respond quickly to novel threats but bounded enough that self is not treated as enemy.

One can think of this as risk management under uncertainty. In finance, the most dangerous portfolios are sometimes those that look safe under normal conditions but fail catastrophically in rare events. Immune systems face an analogous problem. A response strategy optimized for "typical" infections might be brittle to unusual pathogens. Noise and heterogeneity spread risk, ensuring that rare events do not always produce system-wide failure. That is the strategic value of noise: it creates a repertoire not only of receptors, but of behaviors.

There is also a more intimate, personal side to this story. Anyone who has had two colds that felt completely different, or who has watched two family members respond differently to the same infection, has encountered immune noise at the organism level. Some of that variation is genetic and environmental, but some is simply the arithmetic of small numbers. Early in an infection, a handful of infected cells and a handful of responding immune cells can set the trajectory. A slight difference in timing—a cytokine produced a little earlier, a cell arriving a little later—can shift the outcome. The immune system is not deterministic in the way a mechanical clock is deterministic. It is probabilistic in the way weather is probabilistic: governed by laws, yet sensitive to fluctuations.

That probabilistic nature can be unsettling, especially when medicine would like to predict outcomes precisely. Yet it also explains why immune systems can be so adaptable. A deterministic immune system would need an explicit plan for every possible pathogen. A probabilistic immune system can explore multiple response modes in parallel. Some cells try one approach; others try another. If one approach works, feedback and cytokines recruit more cells into that mode. If it fails, the system has not put all its chips on a single number.

The same logic appears in the germinal center, where B cells mutate their antibody genes and compete for survival signals. Mutation is the quintessential biological noise: random changes that are mostly harmful. In the germinal center, mutation becomes productive because selection filters it. Diversity is generated blindly; the environment then rewards the variants that bind antigen better. That is noise disciplined by competition, a microscopic version of evolution happening within days. It is hard to imagine a clearer example of randomness turned into learning.

Noise, then, is not a single thing to be eliminated. It is a family of variations to be shaped. Some noise is suppressed by kinetic proofreading and integrative thresholds, because fleeting signals should not count. Some noise is tolerated because it is harmless. Some noise is actively used because it creates options and spreads risk. The immune system survives by being neither perfectly orderly nor chaotically reactive, but by maintaining a controlled roughness, a texture of variability that can be smoothed when peace returns and roughened when danger demands improvisation.

In the end, what looks like messiness is often a form of preparedness. A pathogen evolves. A tissue changes. A barrier breaks. A new antigen appears. In that shifting landscape, the immune system's most reliable asset may be its refusal to behave like a single, perfectly tuned machine. It behaves, instead, like a population of slightly different machines, each prone to small errors, whose collective behavior can still find the right answer often enough for a body to endure.

Chapter 6

The Delicate Balance of Self

The immune arsenal is potent enough to liquefy organs, yet it generally leaves our own tissues unscathed. We explore the rigorous, often lethal training regime that lymphocytes endure to distinguish 'self' from 'other', and the specialized peacekeeping circuits that actively suppress mutiny when the initial screening fails.

6.1 The Brutal Education of a Lymphocyte

A newborn T cell begins life with a problem that is both practical and philosophical: it must become exquisitely good at noticing danger, while remaining almost perfectly uninterested in the body that made it. The immune system cannot outsource this judgment to a single master list of enemies. Pathogens mutate; cancers evolve; the microbial world is too inventive. So T cells are equipped with receptors that are generated almost at random, each cell assembling its own unique sensing device by shuffling genetic segments like cards. The payoff is breadth: somewhere in that vast lottery of receptors will be cells capable of recognizing nearly any conceivable molecular pattern. The cost is predictable: many of those receptors will recognize *us*.

Central tolerance is the name for the earliest, most uncompromising attempt to solve this. It takes place primarily in a small organ above the heart, the thymus, whose importance is easy to underestimate until it is missing. In the 1960s, Jacques Miller removed the thymus from newborn mice and watched their immune systems collapse: lymphocyte numbers dwindled, responses to infection faltered, and the animals became immunologically fragile. The thymus was not decorative. It was a training ground. But it was also something more severe than training: it was a filter that destroys most of what it produces. A successful T cell is not simply one that can fight; it is one that has been proven, under pressure, to be safe.

To understand the brutality, it helps to say what a T cell is *for*. T cells do not usually recognize whole microbes the way antibodies can. Instead, they specialize in reading the body's internal bulletin boards. Nearly every cell displays small protein fragments on its surface, each fragment nestled in a molecular holder called an MHC molecule (major histocompatibility complex). These fragments are called peptides. They are constantly generated as proteins are made, used, and broken down. The display is a sample of what is happening inside the cell. If a virus is replicating, viral peptides appear; if a cell is becoming cancerous, abnormal peptides can appear; if a cell is healthy, the peptides are ordinary self. A T cell receptor (TCR) is tuned to recognize a combination: a particular peptide *plus* the particular MHC molecule presenting it. This is why T cells are so powerful and so dangerous. They are built to interpret *self presentation*. Every time a T cell checks a cell's MHC display, it is peering into the self. A mistake here is not a small mistake.

The thymus makes this judgment concrete. In the thymus, immature T cells—thymocytes—are forced to interact with self peptide–MHC complexes in a controlled environment, and their fate depends on what those interactions look like. The thymus is often described as having two broad regions: a cortex on the outside and a medulla toward the center. These are not just architectural features; they are different neighborhoods with different teachers and different exams. Thymocytes arrive as raw recruits that cannot yet do anything useful. They lack a fully formed receptor. They proliferate, rearrange their receptor genes, and eventually express a TCR on their surface. From that point onward, they are tested for two qualities that must be held in tension.

The first test is almost counterintuitive: can the thymocyte recognize *self* at all? This is called positive selection. It does not mean the cell is approved because it is self-reactive. It means the cell is approved because it can interact with the body's MHC molecules. A T cell that cannot bind any self MHC at all is useless, because it would never be triggered by any cell in the body; it would wander like a key that fits no lock. In the cortex, thymic epithelial cells present a sampling of self peptides on MHC. A thymocyte whose receptor cannot make even a weak, fleeting connection fails to receive survival signals. It dies quietly, a kind of cellular fading known as death by neglect. The thymus does not *punish* these cells so much as it refuses to keep them alive.

If that were the only test, the thymus would produce an army primed to read self MHC—and therefore, potentially, to attack self tissues. So the second test is harsher. It asks: does the receptor bind too strongly to self peptide—MHC? Strong binding is a proxy for danger. It suggests that, out in the body, encountering that self peptide could trigger the cell into the kind of activation that leads to inflammation and killing. This is the logic of negative selection. When a thymocyte binds self too tightly, it receives a lethal instruction: commit suicide. The mechanism is apoptosis, a programmed cell death that is orderly and contained, designed to avoid spilling inflammatory contents. In a sense, the thymus is a place where self-recognition is required, but enthusiasm for self is fatal.

The result is a Goldilocks problem. Too little binding and a cell is useless. Too much binding and it is a threat. Only a narrow middle range permits survival.

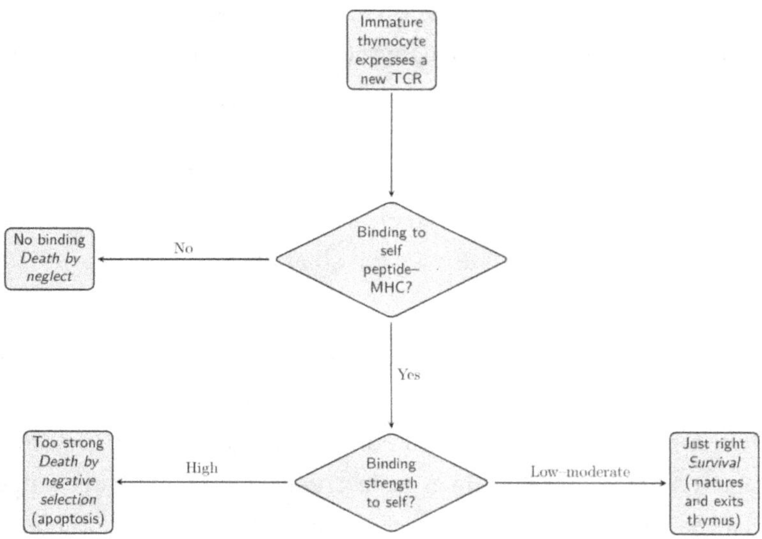

What makes negative selection more than a simple quality-control switch is the challenge of representation. The thymus must show developing T cells a convincing portrait of the body's molecular self. But the body is large and diverse. A protein abundant in the retina is irrelevant to a kidney cell, and vice versa. A protein made only in the pancreatic islets may be rare everywhere else. Many self proteins appear only at certain times in life: during fetal development, puberty, pregnancy, aging, stress. If central tolerance depended on a limited, generic menu of self peptides, it would leave enormous blind spots. Autoimmunity would be inevitable.

The thymus solves part of this with an audacious biological trick: it forces some of its own cells to express proteins that normally belong to distant tissues. Medullary thymic epithelial cells can display "tissue-restricted antigens," a kind of molecular ventriloquism in which the thymus speaks in the voice of skin, liver, pancreas, and brain. The developing thymocyte, encountering these unusual peptides on MHC, is confronted with a broader picture of what counts as self. Dendritic cells in the thymus add another layer. They can capture antigens from surrounding cells and present them, expanding the range of self peptides that thymocytes are tested against. Even so, the portrait is incomplete, and the incompleteness is not a bug so much as a consequence of physics and probability: there are too many possible peptides, too many contexts, too many rare proteins for any single organ to present them all at once.

That imperfection is visible in the clinic. A striking example comes from a rare autoimmune syndrome historically called APECED (autoimmune polyendocrinopathy-candidiasis-ectodermal dystrophy), also known as autoimmune polyendocrine syndrome type 1. Here, a defect in a gene crucial for thymic self-antigen display leads to a peculiar pattern of multi-organ autoimmunity, often involving endocrine glands. The details matter less than the principle: if the thymus cannot convincingly impersonate the periphery, self-reactive T cells that should have been deleted can slip through. This is central tolerance revealed by its failure mode.

The brutality of thymic selection also has a statistical feel. Because receptors are created by random recombination, the thymus must expect enormous waste. Most new receptors will be nonfunctional, or will not bind self MHC at all, or will bind self too well. It is common for the majority of thymocytes to die in the thymus. This sounds extravagant until one remembers the alternative: keeping a dangerous cell. One rogue clone, if activated, can expand into millions of descendants and orchestrate a coordinated attack. Deleting that clone early is cheaper than suppressing its rebellion later.

Yet there is a subtlety in how "too strong" is interpreted. The thymus is not measuring moral guilt; it is measuring signal strength. When a thymocyte's receptor binds a peptide—MHC complex, it triggers a cascade of intracellular signals. The cell integrates how strong the signal is, how long it lasts, and the context in which it occurs. These signals are biochemical, not conscious, but they behave like decisions: thresholds are crossed, gene programs are switched on, survival factors are gained or denied. Central tolerance is often introduced as a simple story—weak survives, strong dies—but the underlying biology is more like a set of competing fates that depend on timing and tissue microenvironment.

One reason is that MHC itself comes in two broad classes, and T cells split into two major lineages that match them. CD8 T cells, sometimes called cytotoxic T cells, are adapted to recognize peptides presented by MHC class I, which is on nearly all nucleated cells and is particularly suited to reporting what is happening inside them. CD4 T cells, often called helper T cells, recognize peptides presented by MHC class II, which is mostly on professional antigen-presenting cells such as dendritic cells, macrophages, and B cells. During thymic development, a thymocyte expresses both CD4 and CD8 for a period (a "double-positive" stage). As it is positively selected, its receptor's preference for class I or class II MHC nudges it toward becoming a CD8 or CD4 cell, respectively. This is another kind of narrowing: not only is the receptor tested for acceptable self-reactivity, but the cell is assigned to a functional role based on what it can see.

The medical consequences of getting this wrong became widely known outside immunology during the HIV/AIDS crisis. HIV targets CD4 T cells, and the loss of these helper cells cripples immune coordination, leaving patients vulnerable to infections that a healthy immune system keeps at bay. That clinical fact underscores the stakes of thymic education. The thymus is not merely preventing autoimmunity; it is sculpting the very categories of cellular labor the immune system relies on.

The notion of "self" also includes a twist that can feel almost paradoxical: a certain degree of self-reactivity is not only tolerated but necessary. Positive selection ensures that mature T cells will react to peptides presented on self MHC at least weakly. Out in the body, this weak self-recognition helps keep T cells alive through tonic signaling—a kind of basal reassurance that they still belong in the system. The immune system, in other words, does not build T cells to ignore self entirely. It builds them to be self-literate without being self-obsessed.

There is also the question of what happens when a thymocyte recognizes self quite strongly but not in a way that leads to outright deletion. For some cells, strong self-recognition can divert them into a special lineage designed to restrain immune responses: regulatory T cells. The existence of this lineage hints at an older truth about biological control: when you cannot eliminate every risk, you sometimes convert part of the risk into a stabilizing force. A cell that recognizes self well is well positioned to patrol places where self is abundant and to intervene when inflammation begins to run away. But diversion into regulation is not guaranteed; it depends on additional signals and on the cellular context in which the self-recognition occurs. The thymus is thus not only a slaughterhouse but also a sorting facility, allocating rare survivors into different kinds of future.

Even the language of "selection" can mislead by implying a fixed exam with clear answers. In reality, the thymus is a changing environment. The peptides displayed depend on which proteins are being expressed, processed, and presented by thymic cells at that time. The density of peptide—MHC complexes, the availability of co-stimulatory molecules, the local cytokines (small signaling proteins), and the movement of thymocytes through thymic regions all influence outcomes. A thymocyte may encounter its most dangerous self peptide only if it migrates to the right place at the right time. Some negative selection likely occurs in multiple thymic niches, and immunologists have long debated the exact geography of where the most consequential deletion happens. What matters for the developing cell is not the map but the arithmetic: the thymus increases the odds that a harmful receptor will be caught before it can leave.

Seen from the outside, the process can sound needlessly cruel. Why generate so many cells only to destroy them? The answer is that the immune system's most valuable property—its ability to anticipate novelty—requires a generator that does not know in advance what will be needed. Randomness is not a side effect; it is a strategy. But randomness produces hazards. Central tolerance is the price paid for daring to be imaginative.

A final, sobering point is that the thymus does not do this work forever with equal vigor. It is most active in early life and gradually involutes with age, shrinking and becoming more fatty. Adults still produce new T cells, but at lower rates, and the immune repertoire becomes more dependent on long-lived memory and on the maintenance of existing clones. This shift helps explain why immune responses change across the lifespan: why children respond differently to new pathogens, why certain autoimmune diseases have characteristic ages of onset, why recovery from immune depletion can be harder later in life. The thymus is not merely a place; it is a phase of immunological possibility.

For a T cell that survives, the reward is not comfort. It is permission to circulate. It leaves the thymus carrying a receptor that has been tested against an imperfect but extensive representation of self, tuned to recognize peptide—MHC combinations without tipping too easily into self-destruction. The cell enters a body filled with the same molecules it has already seen in curated form. That familiarity is not a cue for attack. It is the background against which true danger must stand out, sharply enough to justify the risks of immune violence. In the quiet equilibrium of a healthy person, the success of central tolerance is measured by an absence: the lack of civil conflict, the lack of damage that would occur if even a small fraction of those rejected recruits had been allowed to graduate.

6.2 The Peacekeepers

Even after the thymus has done its ruthless pruning, it would be a mistake to imagine that the body is populated only by perfectly safe T cells. Central tolerance is a powerful filter, not a magical one. Some self-reactive cells slip through because the thymus never displayed the particular self peptide they would have overreacted to, or because the interaction landed just under the threshold for deletion, or because biology is noisy and thresholds are never razor sharp. A mature immune system therefore lives with a permanent, managed risk: cells with the capacity to recognize self circulate in the blood and enter tissues every day. The question is how that risk is kept from becoming injury.

Peripheral tolerance is the name for the second line of restraint. It is not confined to one organ. It is distributed across the body, embedded in the ordinary routines of immune surveillance, and enforced by an ensemble of mechanisms that behave less like a courtroom and more like municipal governance: quiet rules, constant monitoring, and a preference for de-escalation unless danger is unmistakable. Among the most important agents of that governance are regulatory T cells, often abbreviated Tregs. They are peacekeepers in the literal sense: cells whose job is not to win fights, but to prevent fights from starting, and to stop them when they have gone too far.

A T cell that has left the thymus is not automatically a weapon. It is more like an unloaded gun with a complicated safety catch. To become a fully armed effector cell—a helper that orchestrates inflammation or a killer that destroys infected cells—a T cell generally requires at least two kinds of input. The first is recognition: its receptor binds a specific peptide displayed on an MHC molecule. The second is context: a set of additional signals, delivered by antigen-presenting cells, that indicate the peptide is part of a meaningful threat rather than background self. Those additional signals can take many molecular forms, but the underlying idea is simple enough for a dinner table: the immune system is wary of acting on a single clue. A false alarm is costly.

Dendritic cells are central to this contextual decision. They are scouts that pick up material from tissues and carry it to lymph nodes, where T cells gather. When dendritic cells encounter microbes or tissue damage, they switch into an activated state, displaying more co-stimulatory molecules and secreting inflammatory cytokines that tell T cells, in effect, *this matters*. But most of the time, dendritic cells are in a resting or "steady-state" mode. They still present peptides—including many self peptides derived from normal tissue turnover—but they do so without the full set of alarm signals. A self-reactive T cell encountering its antigen in this calm setting is often not activated into a fighter. Instead it can be rendered unresponsive, a state known as anergy, or it may be deleted, or it may be steered toward a regulatory fate. Peripheral tolerance, in other words, is not only suppression of bad behavior; it is also education by boredom. The immune system teaches certain cells that their favorite target is not worth reacting to.

This "context" logic becomes especially vivid when it breaks. Consider checkpoint inhibitor immunotherapy, a modern triumph in cancer treatment that also functions as an accidental experiment in immune restraint. Drugs that block inhibitory receptors such as CTLA-4 or PD-1 can unleash T cells against tumors by removing molecular brakes that ordinarily dampen activation. For many patients, this can lead to dramatic tumor regression. But the same brake removal can also produce inflammation in healthy organs: colitis, hepatitis, thyroiditis, skin rashes, and other immune-related adverse events. The pattern is a clinical reminder that the immune system is not held back by ignorance alone. It is held back by active, continuously applied inhibition. When that inhibition is pharmacologically interrupted, self-reactive cells that were previously contained can become loud and destructive.

To understand how peacekeeping works at the cellular level, it helps to meet the main peacekeepers.

Regulatory T cells are a specialized subset of CD4 T cells. Many of them express a transcription factor called FOXP3, which acts like a master switch for a regulatory program: it alters which genes are turned on and off, shaping the cell into a suppressive, stabilizing influence. The existence of Tregs reveals a deep design principle. The immune system does not rely solely on deleting dangerous clones, because deletion is never complete and because some degree of self-recognition is woven into normal T cell biology. Instead, it maintains a population of cells whose *purpose* is to restrain other immune cells, particularly when the target is self.

The historical roots of this idea have a strangely sharp edge. In 1995, Shimon Sakaguchi and colleagues performed experiments in mice that turned a hazy concept into a concrete mechanism. They worked with CD4 T cells and focused on a subset that expressed CD25, the alpha chain of the IL-2 receptor. When that CD25-positive subset was removed and the remaining T cells were transferred into certain mice, the recipients developed spontaneous autoimmune disease across multiple organs. When the CD25-positive cells were added back, disease could be prevented. The message was stark: the immune system contains a dedicated group of cells that actively prevents self-destruction. Autoimmunity was not simply a failure to delete the wrong clones in the thymus; it could be caused by the loss of a policing population in the periphery.

Human genetics gives the same message with even greater emotional force. Rare children born with mutations in FOXP3 develop a condition called IPEX (immune dysregulation, polyendocrinopathy, enteropathy, X-linked). In its classic form, it appears early in life with severe diarrhea from intestinal inflammation, eczema-like skin disease, and autoimmune attack on endocrine organs such as the pancreas. These are not subtle immune quirks. They are system-wide breakdowns of restraint, often requiring aggressive immunosuppression or bone marrow transplantation. A child with IPEX provides a grim proof of concept: without functional Tregs, the immune system's default is not calm. The system does not merely *fail to stop* autoimmunity; it *actively generates* it.

How do Tregs keep the peace? They do not have a single weapon; they have a toolkit. Some of it is physical. Some of it is chemical. Much of it involves manipulating the same signals that ordinary T cells rely on.

One contact-dependent mechanism involves CTLA-4, an inhibitory receptor that Tregs often express at high levels. Ordinary T cells need co-stimulation through a receptor called CD28, which binds to molecules (often called B7) on antigen-presenting cells. CTLA-4 competes for those same molecules and can effectively strip them from the surface of the antigen-presenting cell, reducing its ability to provide co-stimulation. The result is a quieter conversation between antigen-presenting cells and would-be effector T cells. This is peacekeeping as infrastructure: Tregs adjust the environment so that the ignition keys are harder to turn.

Another mechanism is cytokine-mediated suppression. Cytokines are soluble messenger molecules, and Tregs can release anti-inflammatory cytokines such as IL-10 and TGF-β, which dampen activation and encourage tissue repair programs. These signals do not target only one aggressive clone; they alter the tone of an entire local immune neighborhood, making it harder for inflammation to escalate.

Tregs also influence access to IL-2, a growth factor that activated T cells crave. Because Tregs express high levels of CD25, they can soak up IL-2 in their vicinity, reducing the amount available to fuel the proliferation of effector cells. This is sometimes described as cytokine "consumption," but the deeper point is that immune responses are not only about recognition. They are about resources. A potential rebellion needs supplies.

These mechanisms interact with the inhibitory receptors that act as immune checkpoints, especially PD-1. PD-1 is expressed on activated T cells, and when it engages its ligands (PD-L1 or PD-L2) on other cells, it transmits signals that reduce T cell activity. This pathway is important in preventing overactivation during chronic stimulation and in maintaining tolerance in tissues. It is one of the reasons some self-reactive T cells can persist without causing disease: they are continuously restrained by inhibitory signaling, not merely ignored. Blocking PD-1 can therefore reveal hidden autoimmune tendencies, as seen in checkpoint inhibitor side effects. At the same time, tumors can exploit PD-1/PD-L1 to protect themselves, turning peacekeeping into a hostage situation. The same molecular brake that prevents unnecessary self-harm can be used by cancer to evade elimination.

There is a temptation to treat peripheral tolerance as a simple backup plan: if central tolerance misses something, Tregs catch it. The truth is more interesting and more personal. Peripheral tolerance allows the immune system to adapt restraint to the realities of life, including the fact that "self" is not a static object.

Take the gut. The intestine is a paradox: it is a barrier exposed to an enormous mass of foreign material, much of it harmless or beneficial, and it is also a delicate tissue that cannot afford constant inflammation. The immune system must ignore most food proteins and coexist with trillions of microbes while remaining ready to respond to invasive pathogens. This balance depends heavily on regulatory pathways. It is not enough to delete self-reactive clones in the thymus, because the gut is not merely defending self; it is tolerating non-self that must be treated as functionally permissible. Tregs in the gut, shaped by local cues, help prevent immune responses to diet and commensal microbes. When that balance fails, inflammatory bowel diseases can emerge, and treatments that alter immune checkpoints can tip it further. The gut makes clear that peacekeeping is not the opposite of immunity; it is one of the conditions that makes immunity workable.

Peripheral tolerance also matters in pregnancy. A fetus carries paternal antigens that are foreign to the mother. Yet most pregnancies proceed without the mother's immune system mounting a destructive attack. This is not because the fetus is invisible. It is because multiple tolerogenic mechanisms are active at the maternal-fetal interface, including regulatory pathways that dampen potentially harmful T cell responses. Pregnancy illustrates how the immune system must sometimes choose restraint in the face of genuine antigenic difference, because the cost of aggression would be catastrophic.

A central challenge for any peacekeeping system is discrimination: knowing when to hold back and when to let the fighters fight. Here, again, dendritic cells and context matter. When tissues are healthy, antigen is presented without danger signals, and tolerance is favored. When tissues are infected or damaged, antigen presentation is accompanied by inflammatory cytokines and co-stimulation, and immunity is favored. The immune system does not ask, *Is this molecule self or non-self?* as a single binary question. It asks, *What does this presentation mean right now?* It is a decision made under uncertainty, with feedback and reversible states.

That reversibility is essential. A T cell can be anergic today and dangerous tomorrow if the surrounding signals change. A Treg can be stable and suppressive in one context and less effective in another, especially in the presence of intense inflammation. Even FOXP3, the defining marker of many Tregs, is part of a program that must be maintained. Some regulatory cells are long-lived and committed; others are induced in the periphery from conventional T cells under certain conditions, especially in mucosal tissues. The immune system thus maintains peace through both standing forces and reserve forces, shaped by experience and environment.

The real-world consequences of this distributed system show up in subtler ways than rare genetic syndromes. Many people carry genetic variants that slightly change the thresholds of immune activation or the stability of regulatory programs. Add infections, stress, hormonal changes, microbiome shifts, tissue injury, and chance, and the probability landscape of autoimmunity changes. One person may go through life with a few self-reactive clones kept quietly in check. Another may have a small perturbation—a viral infection that strongly activates antigen-presenting cells, for example—that turns an otherwise contained self-reactive population into an expanding effector response. In that moment, Tregs and checkpoints are not abstract concepts; they are the difference between a fleeting flare and a chronic disease.

There is also a psychological comfort in imagining that "the immune system" is a single entity that either recognizes self or does not. Peripheral tolerance reveals something messier and more human in its logic. It is closer to a society than a machine: there are rules, but also exceptions; enforcement is constant, but not perfect; mistakes can be corrected, but sometimes they cascade. Peace is not the default state granted by ignorance. It is the outcome of active work by cells that spend their lives preventing the immune system's extraordinary power from becoming a liability.

A regulatory T cell, sitting in a lymph node or embedded in a tissue, does not look heroic under a microscope. It is small, unassuming, indistinguishable to a casual glance from many other lymphocytes. Yet its role is quietly profound. It allows the immune system to preserve a repertoire broad enough to recognize the unexpected, without living in constant fear of itself. It makes space for useful self-recognition—the low-level interactions that keep T cells alive and calibrated—while limiting the probability that those interactions ignite. It turns restraint from a passive absence of activation into an active physiological function, maintained moment by moment.

When the system works, the evidence is negative: no rash, no diarrhea, no inflamed joints, no failing glands. The body's tissues go about their business without being treated as suspicious. A scrape heals without spiraling into chronic inflammation. A new meal is digested without an immune argument. A tumor, if it arises, is faced by a system that must decide how hard to press the accelerator without crashing into the guardrails. In that ongoing negotiation, the peacekeepers are not a luxury. They are the terms on which immune memory can exist without turning the body into collateral damage.

6.3 Civil War

Autoimmune disease is often described as the immune system "attacking the body." The phrase is vivid, but it hides the more unsettling truth: much of autoimmunity is not a single wrong target so much as a wrong *decision.* The immune system is built to decide under uncertainty. It must weigh evidence that something is dangerous against evidence that it is normal. It must commit to action before it has perfect information, because pathogens do not wait. Under ordinary conditions, a mesh of safeguards keeps those decisions from tipping into self-harm. When that mesh fails, the resulting conflict can look like a civil war: the same mechanisms that ordinarily protect and repair become sources of chronic damage, and the memory of past battles makes the next skirmish easier to ignite.

A useful way to think about autoimmunity is as a failure of an algorithm. Not an algorithm in a computer, but a set of biochemical rules that determine what happens when a T cell or B cell encounters an antigen—a piece of protein recognized by receptors. Those rules integrate multiple inputs: how strongly the receptor binds, whether the antigen is presented with co-stimulation, what cytokines are present, whether inhibitory receptors are engaged, whether regulatory T cells are nearby, whether the tissue is already inflamed, whether the antigen is persistent. The "output" is a fate decision: activation, silence, deletion, regulation, exhaustion. Autoimmune disease emerges when the outputs drift systematically in the wrong direction.

One can see this drift most clearly by looking at the kinds of mistakes the system is capable of. The first mistake is a failure of representation: the developing immune repertoire was not adequately shown what counts as self. The second is a failure of restraint: self-reactive cells that exist in the periphery are not adequately suppressed. The third is a failure of termination: once an immune response begins, it cannot be shut down efficiently, and the cells that should die instead persist and accumulate. These are not separate diseases so much as different weak points in the same architecture.

A dramatic example of the first mistake is APECED, the syndrome linked to defects in a gene called *AIRE*. *AIRE* helps certain thymic cells display a wider sampling of tissue-specific self proteins during T cell development. When that display is impaired, self-reactive T cells that should have been deleted can mature and exit into the circulation. The clinical picture that follows can involve multiple endocrine organs and other tissues, often with a distinctive mix of autoimmunity and susceptibility to chronic mucocutaneous Candida infection. The details can feel bewildering until the logic clicks: if central tolerance is patchy, the immune system is trained on an incomplete map of the self, leaving some territories effectively unrecognized as friendly. In that situation, even a well-functioning peripheral peacekeeping system may be overloaded, forced to contain a larger set of potentially aggressive clones than it was designed to manage.

The second mistake is made stark by IPEX, caused by mutations in *FOXP3*. Without *FOXP3*, regulatory T cells are absent or dysfunctional, and the brakes on immune activation fail in an especially catastrophic way. Children with classic IPEX can develop severe enteropathy, eczema, and autoimmune endocrinopathies early in life. The violence of the phenotype is instructive: the immune system is not a purely reactive defense force that needs to be switched on; it is an engine that must be actively restrained. Without the restraining program, inflammation becomes a default mode, and self becomes a constant trigger.

The third mistake, failure of termination, can be illustrated by ALPS: autoimmune lymphoproliferative syndrome. Here the key is a pathway centered on FAS, a receptor that helps trigger apoptosis—programmed cell death—in immune cells that have finished their job or that are dangerously self-reactive. A healthy immune response is expansive and then contractive. It grows a population of activated lymphocytes to address a threat, then prunes that population back down. This contraction is not aesthetic tidiness; it is how the system prevents activated cells from lingering long enough to cause mischief. In ALPS, defects in the FAS pathway can lead to lymphoproliferation (an abnormal accumulation of lymphocytes), enlarged lymph nodes and spleen, and autoimmune manifestations. What makes ALPS especially revealing is a feature geneticists call incomplete penetrance: some people with a pathogenic mutation develop disease, while others remain relatively asymptomatic. The mutation is real, but the outcome is probabilistic. It depends on the rest of the immune system's parameters and on life history—infections, tissue injuries, and the many small variables that determine how often the immune system is asked to expand and contract. Civil war becomes more likely when the demobilization machinery is defective, but it is not guaranteed. The immune system is an ecosystem; it can sometimes compensate.

These monogenic syndromes are rare. Their value is that they illuminate principles that also apply, more subtly, to common autoimmune diseases such as type 1 diabetes, multiple sclerosis, rheumatoid arthritis, systemic lupus erythematosus, and inflammatory bowel disease. Most patients do not have a single broken gene that explains everything. Instead, they carry a constellation of risk variants, each nudging thresholds slightly. Those thresholds only matter when events push the system toward a decision point. That is where triggers enter the story.

One of the most discussed triggers is molecular mimicry. The idea is deceptively simple: a pathogen carries a peptide that resembles a self peptide closely enough that immune cells trained to recognize the pathogen may also recognize the self tissue. It is not necessary for the sequences to be identical. The immune receptor does not read a peptide like a sentence; it recognizes a three-dimensional shape and a chemical texture presented in the groove of an MHC molecule. Two different peptides can present a similar "face" to the receptor. If a T cell is activated strongly during infection, it can expand and form memory. If that same receptor also binds a self peptide in, say, the heart, joints, or nervous system, the immune response may continue after the pathogen is gone, now redirected toward self.

A classic clinical narrative often discussed in this context is rheumatic fever, which can follow infection with certain strains of *Streptococcus pyogenes*. The immune response to streptococcal components can, in susceptible individuals, cross-react with heart tissues, contributing to inflammation that can damage valves. Not every patient with strep throat gets rheumatic fever; far from it. The rarity fits the algorithmic framing: the system requires a particular convergence of antigenic resemblance, genetic background (especially in antigen presentation), and immune context. The infection supplies strong co-stimulation and inflammatory cytokines, which are precisely the conditions that favor activation and durable memory. In such a state, the immune system is willing to take risks. If one of the expanding clones has a second allegiance to a self target, the bargain becomes dangerous.

Molecular mimicry is compelling because it provides a plausible *origin story* for how tolerance could be broken. But origin stories can mislead if they imply a single decisive moment. Autoimmunity more often behaves like a feedback loop.

Feedback loops are common in biology because they create persistence. They also create traps. In autoimmunity, a particularly vicious loop can begin with tissue damage, from almost any cause. Damaged cells release intracellular contents and altered proteins.

Local antigen-presenting cells pick up these materials and present them. Inflammation increases co-stimulation and cytokine production. Self-reactive T cells that were previously anergic or restrained may now receive sufficient activation signals. They help activate B cells, which can produce autoantibodies. Those antibodies bind to tissues, activate complement, recruit innate immune cells, and cause more damage. More damage creates more antigen and more inflammatory context. The system learns the wrong lesson: it forms immune memory not of a pathogen, but of an internal pattern associated with injury.

Systemic lupus erythematosus offers a vivid example of how such loops can involve nucleic acids, the cell's DNA and RNA. Normally, the immune system is cautious about responding to nucleic acids because every cell contains them. Yet when cells die in certain ways or when clearance of cellular debris is impaired, DNA- and RNA-containing complexes can be presented to immune sensors and to B cells. Autoantibodies to nuclear components can form immune complexes that deposit in tissues such as the kidney, triggering inflammation that releases more cellular material. The particulars of lupus are complex, but the loop logic is recognizable: self material becomes persistent antigen, and inflammation provides the context that keeps presenting it as dangerous.

The language of "broken feedback loops" is not metaphorical. Immune responses are regulated by negative feedback at many levels: inhibitory receptors such as PD-1 dampen signaling; regulatory T cells suppress activation; anti-inflammatory cytokines reshape the environment; activated cells upregulate pathways that make them more likely to die once stimulation wanes. In autoimmunity, these feedbacks may be insufficient for the magnitude of activation, or they may be partially disabled by genetic variants, or they may be overwhelmed by ongoing antigen availability. The system then behaves like a microphone too close to a speaker. A small input becomes a howl.

This is where immune memory becomes both a triumph and a liability. Memory T cells and memory B cells respond faster and more vigorously upon re-encountering their antigen. That is the whole point of vaccination and of durable protection after infection. In autoimmunity, the “antigen” is not going away. A self protein in the thyroid, or myelin in the nervous system, or insulin-related peptides in pancreatic islets, is not a transient intruder; it is a permanent feature of the body. If the immune system accidentally stores a self-directed memory, it stores a memory of something it will encounter every day. The consequence can be chronicity: flares and remissions, progressive tissue damage, and a tendency for the disease to rekindle after apparent quiet.

An additional layer of complexity comes from a phenomenon called epitope spreading. An epitope is the specific part of an antigen that a receptor recognizes. Early in an autoimmune process, the response might focus on one epitope. As tissue damage continues, other proteins from the same tissue are released and presented, and the immune response broadens to new epitopes. The war expands its front. This helps explain why autoimmune diseases can evolve over time, becoming harder to treat as more immune clones become involved.

Therapies for autoimmunity are often, implicitly, attempts to re-tune the algorithm. Some bluntly reduce immune activation across the board, trading infection risk for reduced tissue damage. Others aim more specifically at particular cytokines or cell types. The success and failure of these therapies also reveal something about causation. If blocking a single inflammatory cytokine dramatically helps a disease, that cytokine was not merely a bystander; it was part of the sustaining loop. If depleting B cells helps, then antibodies or antigen presentation by B cells were likely central. If a therapy works only temporarily, it suggests that the underlying antigenic drive remains and the system has alternative pathways to sustain the response. Autoimmune disease is rarely a single chain of dominos; it is more often a braided rope of interactions.

The same idea is visible in the unintended consequences of therapies that stimulate immunity. Checkpoint blockade, used to treat cancer, can trigger autoimmune-like inflammation because it lifts inhibitory constraints that normally contribute to peripheral tolerance. This is not a moral argument for or against such drugs; it is a mechanistic lesson. The immune system's brakes are not redundancies for a rarely used emergency. They are part of everyday driving. Remove them, and even a competent driver can crash.

Autoimmunity also carries a social and emotional burden that is easy to miss in purely mechanistic accounts. Many autoimmune diseases disproportionately affect women, often striking in early or middle adulthood, when careers, caregiving, and family life are already demanding. Symptoms can be intermittent, invisible, or difficult to explain, which can lead to misdiagnosis or to the exhausting need to justify one's illness. Treatments can be life-changing, but they can also require long-term immunosuppression with its own risks and trade-offs. The immune system's internal civil war is lived externally as uncertainty: not knowing when the next flare will come, or whether fatigue is the disease, the medication, the stress, or all of them together.

Despite all this, it is worth holding onto a surprising point: autoimmunity is not evidence that the immune system is poorly designed. It is evidence that the immune system is designed for a world where the cost of missing a real infection can be fatal. The system errs on the side of responsiveness, and it compensates for that bias by layering restraints. When the layers line up, tolerance is robust. When they misalign—a gap in central tolerance, a weak regulatory program, an unusually inflammatory infection, a defect in apoptosis, a tissue injury that does not resolve cleanly—the same learning and memory that usually save lives can turn inward.

The image of civil war captures not only the damage but also the tragedy of misapplied skill. The immune system brings to bear exquisite specificity, rapid mobilization, and long-term memory. It is capable of coordination across tissues and years. In autoimmune disease, those abilities do not disappear; they are redirected. The immune system keeps careful records of the wrong enemy, reinforces the wrong associations, and treats familiar tissues as recurring threats. Restoring peace, then, is not a matter of simply suppressing violence. It is a matter of changing what the system believes, at a molecular level, about what is worth fighting.

Chapter 7

Hacking the System

Vaccination is often viewed merely as a medical shield, but biologically, it is a sophisticated informational hack. By simulating specific patterns of infection, we trick the immune system's decision-making algorithms into registering a threat where none exists, thereby encoding a memory of a war that never happened.

7.1 The Milkmaid's Paradox

A strange bargain sits at the root of vaccination: safety bought with resemblance. Not by meeting the enemy head-on, but by colliding—more gently—with a close cousin. Catch the mild thing, and the severe thing may never take hold. It sounds like folk wisdom because, for a long time, it was. Yet the idea contains a deep immunological truth: the immune system does not need a perfect portrait of a pathogen to remember it. A convincing sketch can be enough.

To appreciate how radical this was, it helps to picture what people feared. Smallpox was not an abstract menace; it was a presence in family albums and graveyards. Survivors could be left with pitted scars, blindness, or both. Epidemics did not simply raise the death rate; they rearranged social life. A disease that could strike children, disfigure the living, and spread through ordinary contact forced communities into a constant calculation: when to gather, whom to visit, whether to care for the sick up close or from a doorway.

Long before anyone knew about viruses, immune cells, or antibodies, communities still noticed patterns. One of the most consequential patterns was that surviving smallpox seemed to grant long-lasting protection. People who had "had it" could nurse the ill with less fear. In modern terms, their immune systems carried memory. In the old terms, they were "safe." Once that observation is in place, a dangerous possibility follows: if the disease itself grants protection, could one obtain the protection without paying the full price?

That question produced one of history's most unsettling public health practices: variolation. Material from a smallpox lesion-dried scabs or pus-was introduced into the skin or nose of someone who had not been infected. The aim was to cause a controlled case. It was not magic; it was a wager. Many variolated people developed a milder illness than they would have acquired naturally, and many were protected afterward. But some died, and some became sources of outbreaks. The practice spread anyway, because smallpox was worse. When the baseline risk is catastrophic, even a risky intervention can look like mercy.

Variolation already contained the essence of vaccination: deliberately exposing the immune system to something in order to teach it. The "milkmaid's paradox" begins when the teaching material changes. Instead of using the deadly pathogen itself, could one use something similar enough to train the body, but too weak—or too wrong, biologically—to do serious harm?

The story often gets told with a single sparkling anecdote: a milkmaid, proud of her clear skin, remarks that she cannot get smallpox because she has had cowpox. It is a good line. It flatters common sense and makes scientific discovery sound like simply listening to the right person at the right time. Historians, however, have complicated the tale. The connection between cowpox and protection from smallpox appears to have been part of rural knowledge in parts of England, and it surfaced through multiple people and moments rather than a single epiphany. The legend persists because it compresses a messy social process into one memorable sentence.

Even if the line was polished after the fact, the underlying observation was real enough to change the world. Cowpox was a disease of cattle that could sometimes infect humans who worked closely with them, especially milkers. In people, it tended to cause localized lesions—unpleasant, sometimes painful, but rarely life-threatening. Smallpox, by contrast, spread efficiently between humans and could be lethal. The paradox was that a brush with the animal disease seemed to purchase protection against the human one.

What could "protection" mean here, if no one had ever heard of a lymphocyte? It meant that after cowpox, the body was altered in a durable way. Exposure left behind something like a memory—an internal readiness that could be called upon later. Yet the protective memory did not require encountering smallpox itself. A related infection was enough.

This is the first glimpse of an idea that now sits at the center of immunology: the immune system recognizes *features*, not names. A pathogen is not one indivisible object to the immune system; it is a collection of molecular shapes. Some of those shapes are shared across related organisms. If the immune system learns to recognize a shared feature, it can respond quickly when a different organism displays the same feature. The teaching does not have to be identical to the exam.

In modern language, those recognizable features are called *antigens*, and the specific parts of an antigen that immune receptors bind are called *epitopes*. These terms can sound clinical, but the underlying concept is familiar. A face is an "antigen" to your visual system; certain features—an eyebrow shape, a smile line—are "epitopes" you rely on for recognition. You do not need a perfect photograph to recognize a friend across a room. A silhouette can suffice if it carries the right cues. Immunity works with a similar economy: it generalizes from what it has seen to what it might soon face, as long as the overlap is meaningful.

The poxviruses provide unusually good material for this kind of generalization. Cowpox virus and variola virus (the cause of smallpox) are distinct, but related enough that their proteins share epitopes. An immune response provoked by cowpox can generate antibodies and T cells that cross-react with smallpox—meaning the same immune tools can bind and attack both viruses. Cross-reactivity can be a blessing or a curse. Here it was a blessing on a planetary scale.

Before the famous London physicians and Royal Society correspondence, the crucial experiment was often domestic and risky. Benjamin Jesty, a Dorset farmer, is recorded as having deliberately inoculated his family with cowpox in 1774, decades before cowpox vaccination became formally established. One can imagine the scene without romanticizing it: a man weighing the dangers of smallpox against the uncertainties of a procedure that respectable medicine had not endorsed. The act was both practical and socially perilous. An experiment carried out on one's wife and children is not a laboratory trial; it is an expression of dread. It also exposes how scientific credit can be shaped by status and publication. Quiet successes in the countryside do not automatically become accepted practice.

By the 1790s, Edward Jenner, a physician in Gloucestershire, took this scattered knowledge and tried to anchor it in a form that could travel: a repeatable procedure with a written rationale. In 1796 he inoculated a boy, James Phipps, with material from a cowpox lesion, and later exposed him to smallpox by variolation. Phipps did not develop smallpox. The experiment, judged by modern ethics, is alarming. Judged by the context of the time, it was a deliberate attempt to replace a risk with a lesser one, supported by an observation that seemed robust. Jenner's real innovation was not merely performing an inoculation, but insisting that it could be standardized, reported, and disseminated as a safer alternative to variolation.

The word *vaccine* itself preserves the animal in the method: *vacca*, Latin for cow. It is worth pausing over that, because it captures something psychologically important. A disease of humans, feared for centuries, was "hacked" using an animal illness. This is not the usual direction of human pride. It suggests humility before biology: the recognition that our bodies can be trained by what looks, at first glance, like a crude trick.

What, precisely, made it a trick worth trusting? The immune system's memory must satisfy two conditions to be useful. First, it must be specific enough to avoid attacking everything indiscriminately. Second, it must be broad enough to handle variation. Pathogens mutate. Even without mutation, no two infections are identical. If immune memory were as literal as a wax seal, it would fail as soon as the pathogen changed a notch. If it were too vague, it would produce chronic false alarms.

Cowpox vaccination succeeded because it threaded that needle. It was different enough from smallpox to be far safer, but similar enough to imprint the immune system with a set of recognition tools that could be redeployed. The paradox is that the immune system can be both picky and forgiving: picky about molecular shapes, forgiving about the source.

There is another subtlety. For protection to be durable, the immune system has to commit resources. It has to create long-lived memory B cells (which can rapidly generate antibody-producing cells on re-exposure), long-lived plasma cells (which can continuously secrete antibodies for years), and memory T cells (which coordinate and kill infected cells). These are not free. Maintaining a vigilant standing army costs energy and space. The body therefore behaves like a careful accountant: it does not usually build durable memory in response to every molecular novelty it encounters. It prefers to wait until an encounter feels consequential.

That is why the cowpox trick is more than a matter of resemblance; it is also a matter of *context.* A real infection, even a mild one, comes with a noisy signature: inflammation, tissue disruption, cellular distress signals, and microbial patterns that the immune system has evolved to treat as urgent. A purified protein floating quietly in the bloodstream is, by contrast, easy to ignore. Cowpox, as a live infection in the skin, arrived with the kind of danger-laced context that persuades the immune system to remember. That "persuasion" is not conscious, but it is a decision in the strict biological sense: a set of molecular switches that determine whether the adaptive response becomes a brief flare or a lasting imprint.

In practical terms, early vaccination achieved two things at once. It presented the immune system with a repertoire of pox-like epitopes, and it did so in a way that resembled a genuine infection closely enough to trigger commitment. One could say that cowpox was an *honest liar*: it lied about the threat's identity while honestly conveying the urgency of the threat category.

The benefits were not immediate everywhere. Vaccination had to compete with variolation, with distrust, with logistical problems (keeping infectious material viable, transmitting it safely from person to person), and with the simple inertia of habit. Early vaccination sometimes used material that was not truly cowpox but related viruses, and the biological details were not always clean. Yet even messy implementations can succeed when the underlying principle is strong.

A helpful way to see the shift is to compare the two wagers.

Variolation: "I will give you a controlled dose of the real danger and hope you survive the milder version."

Vaccination (in its first incarnation): "I will give you a different, milder danger that resembles the real one enough to train you."

Both are attempts to bargain with the immune system's ability to learn. The second bargain is more clever because it separates the *lesson* from the *cost*.

The idea that protection can be purchased with a relative, rather than the exact disease, echoes beyond poxviruses. It is visible whenever one infection changes susceptibility to another, for better or worse. The immune system's generalizations can protect, but they can also mislead. Cross-reactivity can sometimes produce partial protection and milder disease, and in other cases it can bias responses in unhelpful directions. The milkmaid's paradox is the benign version: a near miss that saves lives.

When this kind of cross-protection works well, it does something psychologically profound. It turns an invisible internal process—immune memory—into something you can plan around. Protection becomes something that can be *made*, not merely hoped for. That change in mindset is a hinge in the history of medicine. It makes prevention an engineering project.

There is a final irony in the milkmaid story, whether or not a specific milkmaid ever said the famous line. The story is often told as if it were about a simple observation: milkmaids do not get smallpox. But its power lies in the leap from observation to intervention. Many people can notice a pattern. Fewer will dare to test whether the pattern can be converted into a tool. The leap required not only curiosity, but also a willingness to accept that the body can be educated with a substitute.

That substitute did not need to be a miniature smallpox. It did not need to be a diluted smallpox. It only needed to carry the right molecular hints and arrive with the right biological urgency. Once that is understood, vaccination stops looking like a historical quirk and starts looking like a general strategy: identify what the immune system can learn from, then provide it in the safest believable form.

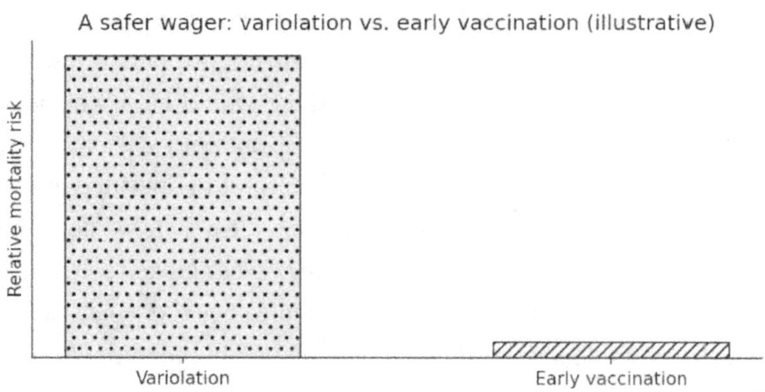

If the immune system could speak, it would not say, "Show me the whole virus." It would say something more pragmatic: "Show me enough to recognize trouble, and do it in a way that convinces me trouble is real." The cowpox wager worked because it satisfied both demands. It gave the immune system a rehearsal that was close enough to the performance to matter, but mercifully far enough away that most people could walk out of the theater alive.

7.2 The Dirty Little Secret

The cowpox trick worked not only because it *resembled* smallpox, but because it arrived with all the noise and mess of an infection: damaged cells, swelling, heat, unfamiliar molecules, and a local alarm that pulled immune cells into the scene. That detail matters because it exposes a puzzle that bothered vaccinologists for generations. If the immune system can learn from exposure, why not teach it with the cleanest possible lesson—one purified protein from a pathogen, refined and sterile, stripped of all risk?

Again and again, the elegant idea failed in practice. The purified protein would enter the body and vanish like a polite guest who never raises their voice. Sometimes there would be a whisper of antibodies. Often there would be almost nothing. Worse, on occasion the immune system seemed to interpret the purified antigen not as a warning but as a cue for *tolerance*—a state of deliberate unresponsiveness. This mismatch between what looked sensible on paper and what actually happened in bodies became what Charles Janeway, with characteristic bluntness, called immunology's "dirty little secret": vaccines often worked because they were *dirty*. They contained microbial contaminants, irritants, or deliberately added substances that made the immune system take the antigen seriously.

The word *adjuvant* comes from Latin for "to help." In vaccines, adjuvants are helper ingredients that amplify and shape immune responses. Many people think of a vaccine as "the antigen," the specific molecular target that the immune system will learn to recognize. Adjuvants remind us that recognition is only half the story. The immune system has to decide that a recognized thing is worth fighting, and worth remembering. That decision is not moral or thoughtful; it is chemical. But it is still a decision in the sense that different inputs lead to different outcomes.

A good way to approach this is to think about what the immune system is trying to avoid. An immune system that attacked everything foreign would be a disaster. Food proteins would trigger constant inflammation. The trillions of bacteria living peacefully in the gut would be treated as invaders. Even harmless pollen would provoke a war. A functioning immune system must be able to *ignore* most of what it encounters. It must also be able to learn "do not overreact" as readily as it learns "attack quickly." In other words, immune memory is not just a record of enemies; it is a record of judgments.

Those judgments are made by cooperation between two arms of immunity. The *innate* immune system is the rapid, ancient part: cells and proteins that respond within minutes to hours, using receptors that are inherited rather than learned. The *adaptive* immune system is slower but exquisitely specific: B cells make antibodies; T cells coordinate and kill; both can form long-lasting memory. If the adaptive system is the part that "remembers," the innate system is often the part that decides what is memorable.

This is clearest in the case of T cells. A T cell does not simply latch onto a free-floating antigen and become activated. It needs a professional presenter: a cell whose job is to sample the environment, pick up proteins, chop them into fragments, and display those fragments like a set of identification cards. The most important of these presenters is the dendritic cell, a sentry that patrols tissues and then travels to lymph nodes to brief the adaptive immune system. Dendritic cells are not impressed by purity. They are impressed by trouble.

T cells, when they first encounter their target, require at least two messages. Immunologists often call them *Signal 1* and *Signal 2*. Signal 1 is the antigen-specific recognition: the T cell receptor binds a particular peptide fragment displayed on a molecule called MHC (major histocompatibility complex) on the surface of the dendritic cell. Signal 2 is the co-stimulatory confirmation: additional molecular "handshakes" and cytokines that indicate the dendritic cell has been activated by danger and is not merely presenting something benign.

Without Signal 2, Signal 1 can be worse than useless. It can teach the T cell that the antigen is not worth responding to. The T cell may become *anergic* (functionally paralyzed) or may be steered into a regulatory role that dampens future responses. This is not a glitch; it is a safety feature. It prevents random encounters with innocuous proteins from generating self-destructive immunity.

The dirty little secret is that many early successful vaccines inadvertently supplied Signal 2. Whole pathogens, even killed ones, bring along molecular patterns that innate immune receptors interpret as "microbe." Even impurities from manufacturing—bits of bacterial cell wall, stray nucleic acids—can supply the missing context. Once vaccinologists began to purify antigens, they often stripped away the very cues that persuade dendritic cells to activate, travel, and instruct T cells.

The immune system's "danger" cues come in two main flavors. One flavor is microbial: patterns that are common in pathogens and uncommon in our own cells. These are called PAMPs, pathogen-associated molecular patterns. They are sensed by pattern recognition receptors (PRRs) such as Toll-like receptors, a family whose importance was cemented when Ruslan Medzhitov and colleagues (1997) showed that a human homologue of the Drosophila Toll protein could trigger pathways that activate immunity. PRRs are not looking for a specific pathogen's name tag. They are looking for motifs that biology has learned to associate with infection: certain types of RNA, particular arrangements of bacterial lipids, unmethylated DNA motifs, and other signatures that are hard for microbes to hide without paying a cost.

The other flavor is the body's own distress signals. When cells are injured, die abruptly, or spill their contents in the wrong place, they release DAMPs, danger-associated molecular patterns. These are not "foreign," but they are informative: they tell the immune system that something is tearing tissue apart. Infection is not the only cause of such damage, but damage is a reliable reason to pay attention.

Adjuvants work by manufacturing context. They give the immune system a reason to treat an antigen as urgent. Some adjuvants mimic microbial patterns and directly engage PRRs. Some create local inflammation or cell stress that releases DAMPs. Many do both. The details vary, and the field has not always agreed on mechanisms, but the functional effect is consistent: adjuvants recruit innate cells, activate dendritic cells, and generate the cytokine environment that determines what kind of adaptive response is built.

Cytokines are small signaling proteins—chemical sentences immune cells use to coordinate. They can promote inflammation, calm it, bias T cells toward different "helper" identities, and influence how B cells refine antibodies. The same antigen presented in two different cytokine environments can produce very different results. One context yields a response dominated by antibodies. Another yields stronger killer T cells. Another yields a muted, tolerant state. This is why vaccines are not merely about *what* you show the immune system, but *how* you show it.

A concrete example of the mystery—and the controversy—surrounding adjuvants is alum, a family of aluminum salts that has been used for decades in human vaccines. Alum is a workhorse: effective, broadly safe, and excellent at boosting antibody responses. For a long time, alum was used without a clear account of how it worked. In 2008, a line of research proposed that alum's adjuvant effects were mediated through the NLRP3 inflammasome, a molecular complex inside cells that helps activate inflammatory cytokines such as IL-1 family members. The inflammasome story was appealing: it provided a mechanistic bridge from a gritty mineral salt to a well-defined innate pathway.

Then other studies complicated the picture, finding that antibody responses to alum-adjuvanted vaccines could persist even when components of that pathway were missing in experimental systems, suggesting that alum's helpfulness might not depend on a single switch. It may act through redundant routes: by causing local cell stress, by altering antigen uptake, by changing how long antigen lingers at the injection site, by shaping recruitment of immune cells, or by engaging different sensors depending on context. This is not an embarrassment; it is a clue. The immune system evolved under conditions where a single point of failure would be dangerous. Redundancy is the rule. A successful adjuvant may be successful precisely because it can tug on several ropes at once.

These ropes lead back to dendritic cells. A resting dendritic cell is like a cautious journalist: it collects information but does not publish a headline. When PRRs are engaged and danger signals are present, the dendritic cell changes character. It upregulates co-stimulatory molecules, processes antigen more aggressively, secretes cytokines, and migrates to the lymph node. There it encounters naïve T cells—T cells that have not yet chosen what they will become.

At this moment, the two-signal logic becomes decisive. Signal 1 says: *this is the target.* Signal 2 says: *this target matters now.* Together they allow the T cell to proliferate, differentiate, and form memory. Without Signal 2, the immune system is effectively saying: *I recognize this, but I'm choosing not to escalate.*

The two-signal idea also resolves a paradox that worries anyone who thinks carefully about immunity: how does the body avoid attacking itself when T cell receptors are generated with random variation?

The answer is not perfect deletion of all self-reactive cells; it is layered control. Some self-reactive cells are eliminated during development, but some slip through. They are kept in check by the absence of danger context and by regulatory systems that promote tolerance. When that control fails—when tissues are inflamed, when infections provide strong innate activation, or when checkpoints are disrupted—autoimmunity becomes more likely. The immune system's dependence on danger signals is a way of lowering the probability that self-recognition becomes self-destruction.

A historical anecdote makes the point in a surprisingly human way. Early vaccine makers often had to keep vaccine material "alive" through serial transfer from arm to arm. In practice, this could mean taking fluid from a vaccine pustule on one person and using it to vaccinate the next. It was an improvised supply chain, vulnerable to contamination and variation. From a modern manufacturing perspective, it is horrifying. From an immunological perspective, it is revealing: those preparations were *unavoidably* full of innate triggers. They were not only delivering antigen; they were delivering a package of inflammatory cues. When the field later sought purity and control, it lost some of the brute-force effectiveness that came from that crude mess.

This is the heart of Janeway's provocation. The immune system is not simply a sensor of *foreignness*. Foreignness is cheap; the body meets foreign molecules constantly. The immune system is a sensor of *meaning*: patterns that imply infection, invasion, or tissue harm. Vaccines succeed when they convey meaning in a way that the immune system cannot responsibly ignore.

One might ask why the antigen itself cannot provide that meaning. After all, if an antigen is truly from a dangerous pathogen, isn't that enough? The immune system does not have access to the category "dangerous pathogen" in the abstract. It has access to molecular patterns and to context. A purified protein from a deadly virus may look, to innate sensors, like any other protein. If it arrives without microbial patterns, without cell stress, without inflammation, it is indistinguishable from a harmless environmental molecule. The immune system's caution is rational: if it attacked every novel protein robustly, life would be one continuous allergy.

Adjuvants therefore function as punctuation. They turn a molecular noun into an urgent sentence. They also shape the tone. Different adjuvants bias the immune response toward different flavors of helper T cells, and those helper T cells, in turn, help B cells make different kinds of antibodies. Some responses are good at neutralizing viruses in the bloodstream. Others are better at recruiting cells to mucosal surfaces. Others are built for confronting intracellular pathogens where killer T cells are needed. A vaccine is not a generic "boost"; it is a carefully biased lesson plan.

The public often hears about adjuvants in the context of anxiety: Why add extra ingredients at all? The immunological answer is simple: because the immune system is designed to be hard to persuade. That stubbornness is part of what keeps us from treating every meal as an emergency. When adjuvants are chosen well, they borrow the language of infection without bringing the full danger of infection.

The two-signal logic also explains why boosters are sometimes needed. If the first exposure produces weak activation—perhaps because innate activation was modest, or because antigen levels were low—the immune system may build only a small pool of memory cells. A booster can expand that pool, increase antibody levels, and improve quality through a process called affinity maturation, in which B cells evolve toward tighter binding. But even boosters rely on the same principle: the immune system must be convinced that the antigen matters enough to be worth revisiting.

Placed alongside the cowpox story, the dirty little secret changes the emotional tone of vaccination history. The earliest vaccines succeeded partly because they were biological impersonations of real infections. Later vaccines had to become more sophisticated because safety demanded that we remove the dangerous parts of infection. Once you remove them, you must replace the missing context deliberately. That replacement is what adjuvants—and more broadly, innate immune activation—provide.

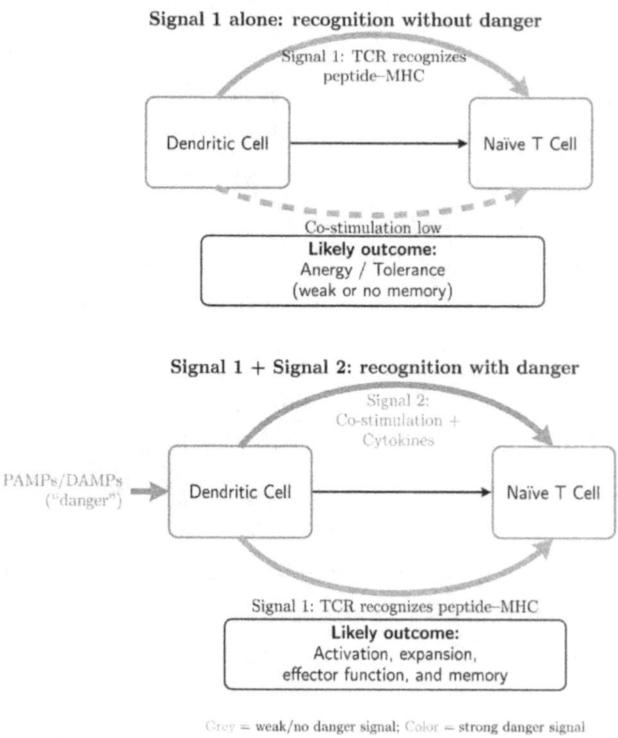

There is a quiet comfort in the fact that the immune system insists on a second opinion. It makes vaccination harder, but it makes ordinary life possible. The body does not hand out lifelong immune grudges to every molecule that happens to pass by. It waits for evidence that something is not merely unfamiliar but consequential, and then it commits—building cells that remember, antibodies that persist, and a readiness that can be summoned at frightening speed the next time the same pattern appears with the same hint of danger.

7.3 The Genetic Instruction Set

Once you accept that the immune system is not impressed by purity, a second question quickly follows. If an effective vaccine must feel, in certain ways, like an infection, why not simply use live pathogens all the time—nature's own teaching tool? The answer is written into every intensive care unit: infection is a brutal instructor. Live-attenuated vaccines, made from weakened viruses or bacteria, can be extraordinarily effective because they reproduce some of the tempo and texture of real disease. But they are not appropriate for everyone, and they carry risks that become ethically hard to justify when safer methods exist.

For much of vaccine history, the compromise was material and mechanical. Take the pathogen, kill it, or break it apart, and inject the pieces. The immune system sees the target, but without replication the encounter can be brief and faint. Add an adjuvant to supply danger context, and the lesson becomes more memorable. This approach has saved countless lives. Yet it still has a limitation that is easy to miss until you look closely at what your immune system is doing during a natural infection.

Infections happen from the inside out. Many pathogens do not merely float through the bloodstream waiting to be neutralized. Viruses, in particular, are obligate squatters: they enter cells and force those cells to manufacture viral proteins. This produces a specific kind of alarm. Infected cells display fragments of internal proteins on their surface, using a pathway centered on MHC class I molecules. That display is how killer T cells—CD8 T cells—discover which cells are compromised and must be destroyed for the sake of the whole organism. Antibodies are superb at blocking viruses before they enter cells, but once a virus is established inside, antibody alone is often not enough. Clearing the infection usually requires a cellular response: T cells that can identify and remove infected cells.

Traditional vaccines can generate some T cell responses, but the most robust CD8 responses tend to arise when antigen is produced within cells, mimicking infection. That is the opening through which genetic vaccines arrived. Instead of injecting a pathogen or its proteins, deliver the *instructions* for making a key viral protein—just enough to stage a rehearsal inside the body, without bringing in the full pathogen.

This sounds like a modern idea, and it is, but it rests on an old intuition: the immune system learns best from events that unfold over time. A bolus of protein injected into muscle is like showing the immune system a snapshot. A cell that produces antigen for a day or two is like running a short film. The immune system is sensitive not only to what it sees, but to the rhythm of seeing: how long antigen persists, where it appears, and whether the context suggests an active invasion.

The simplest version of the genetic instruction set is DNA: a plasmid that encodes an antigen. Cells that take up the DNA can produce the antigen. This approach has been explored for decades and has been used widely in veterinary vaccines, where the practical constraints and regulatory pathways can differ from those for human medicine. DNA vaccination, however, faced challenges in humans, including the efficiency of delivery into cells and robust expression.

mRNA changed the engineering landscape. The logic is blunt: you do not need to deliver a stable genetic library (DNA) to the nucleus if what you want is a temporary script (mRNA) that can be read in the cytoplasm. mRNA is translated by ribosomes into protein, and then it is degraded. It does not need to enter the cell nucleus to work. It does not need to be copied. Its evanescence is part of its appeal. The body receives a set of instructions that are designed to self-destruct after the rehearsal.

The idea that such a fragile molecule could be used as a medicine sounds, at first, like an absurdity. mRNA is easily broken down by enzymes that are everywhere: on skin, in the environment, in tissues. That fragility is one reason biology uses RNA so freely: it can be made and destroyed quickly. Turning that into a vaccine platform required solving two problems at once. First, how do you get mRNA into cells without it being destroyed? Second, how do you make sure the mRNA provokes the right kind of immune response—enough innate activation to teach, but not so much that the body shuts down translation or causes excessive inflammation?

The answer to the first problem, in many modern mRNA vaccines, is a kind of molecular packaging: lipid nanoparticles. These are tiny spheres made of lipids that can encapsulate mRNA, protect it from degradation, and fuse with cell membranes or be taken up by endocytosis. The nanoparticle is not just a shipping container; it is also part of the biological message. Its components can activate innate sensors and contribute to reactogenicity—the soreness, fever, and fatigue that many people interpret as "the vaccine is working," because those sensations are the body's own inflammatory punctuation.

The answer to the second problem required a delicate truce with the innate immune system. Cells have evolved to recognize foreign RNA, because viral genomes and replication intermediates can look like RNA in the wrong place. Pattern recognition receptors—some in endosomes, some in the cytoplasm—detect certain RNA features and trigger antiviral defenses. Those defenses include interferons, powerful cytokines that warn neighboring cells and can slow down protein production. From the immune system's point of view, this is wise: if you suspect viral invasion, you want to slam the brakes on translation. From a vaccinologist's point of view, it is a threat: if cells stop translating, they will not make the antigen, and the adaptive immune system will have little to learn from.

One of the key engineering insights in the mRNA field was that the molecule itself can be tuned. Natural mRNA contains nucleosides—A, U, C, and G—that innate sensors have learned to recognize in certain contexts. By modifying nucleosides, and by optimizing the mRNA's structure (its cap, its untranslated regions, its codon usage), it becomes possible to reduce unwanted innate detection while maintaining robust protein production. The immune response can then be shaped by the delivery system and by the overall context rather than by an overwhelming interferon storm that silences the vaccine's own message.

At first glance, this might sound like trying to hide from the immune system, which seems backwards for a vaccine. But the goal is not invisibility; it is *legibility*. Too quiet, and the immune system ignores the lesson. Too loud, and the lesson cannot even be delivered because the cells stop reading the script. The best vaccines strike a balance between clarity of antigen expression and a controlled level of innate activation that licenses adaptive memory.

The result is a platform that changes what "antigen" means in practice. In older protein vaccines, the antigen is a molecule made in a factory and injected as a finished object. In mRNA vaccines, the antigen is an event: a transient phase in which the body's own cells become antigen factories. That difference has consequences for how antigens are presented to T cells.

When a cell produces a protein internally, fragments of that protein can be displayed on MHC class I molecules, which efficiently engage CD8 T cells. Some of the antigen can also be taken up by antigen-presenting cells and displayed on MHC class II molecules, engaging CD4 helper T cells. Those helper T cells are essential for strong antibody responses because they provide "help" to B cells in germinal centers—the specialized structures in lymph nodes where B cells mutate and compete, improving their antibodies' affinity. In other words, intracellular production can, in principle, feed both sides of adaptive immunity: the antibody arm and the cellular arm.

Viral vector vaccines achieve a similar goal by using a harmless or modified virus as a delivery vehicle for genetic instructions. The vector infects cells, delivers DNA encoding the antigen, and uses cellular machinery to produce it. This can produce strong immune responses, but it also introduces a separate immunological variable: the immune system may respond to the vector itself. Pre-existing immunity to the vector can reduce effectiveness, and repeated dosing can be complicated by anti-vector antibodies. mRNA avoids some of these issues, although it comes with its own constraints in stability, storage, and reactogenicity.

The most vivid historical anecdote in the story of genetic vaccination is also uncomfortably recent, because it took place in real time on screens around the world. When SARS-CoV-2 emerged in late 2019 and spread globally in early 2020, vaccine development faced an old problem at modern speed. Traditional vaccine approaches often require cultivating large amounts of virus, inactivating it safely, or producing protein at scale—processes that can take months to years even under intense effort. Genetic platforms offered a different timeline. Once the viral genome was published, an antigen sequence—most prominently the spike protein—could be selected and encoded rapidly. The work still required careful testing for safety and efficacy, but the design phase became more like software development than like farming.

For the public, this was bewildering. Many people understandably conflated "developed quickly" with "developed carelessly." Yet speed, in this case, came partly from the platform's modularity. If the delivery system and manufacturing process are established, changing the antigen can be as straightforward as swapping the code. The immune system does not care whether the instructions were printed last week or last decade; it cares whether the antigen is expressed and whether the context convinces it to learn.

That modularity also opens the door to rapid updating when pathogens evolve. Influenza vaccines have long been updated to match circulating strains, but the manufacturing pipeline is comparatively slow. Genetic platforms promise quicker iteration. This is not a magic wand—regulatory review, manufacturing capacity, distribution, and public trust still set real limits—but it changes the basic geometry of the problem. Instead of rebuilding a factory line for each antigen, you refine a general-purpose printer.

Another consequence of genetic vaccination is that it clarifies, almost brutally, the difference between *information* and *material*. In an inactivated vaccine, the antigen is matter: a protein, a viral particle, a physical object delivered from outside. In an mRNA vaccine, the payload is information: a sequence that cells interpret to create matter internally. This reframes what a vaccine can be. It can be a set of instructions that causes the body to temporarily mimic a narrow slice of infection—antigen production and presentation—without introducing the pathogen's full destructive repertoire.

There is, however, no free lunch. The immune system treats the act of intracellular antigen production as suspicious, and that is the point. But suspicion has side effects. Many vaccine side effects are, in fact, signs of innate immune activation: fever, malaise, muscle aches. These are not the adaptive immune system at work; they are the body's early warning network throwing cytokine signals around, adjusting metabolism, and recruiting cells. The experience can be unpleasant, and for a small number of people, more serious adverse events can occur, as with any medical intervention. Safety is a continuous negotiation between biological potency and biological restraint.

A deeper, subtler trade-off lies in the kinds of memory that are formed. Antibodies can be long-lived, especially when long-lived plasma cells take up residence in the bone marrow. But antibody levels can wane over time, and pathogens can mutate to evade them. Memory B cells and T cells can persist even when antibodies decline, providing faster and better responses upon re-exposure. Different vaccine platforms and adjuvants bias the balance between immediate antibody titers and durable cellular memory. Genetic vaccines are not automatically "better" at everything; they are another way to sculpt the immune system's choices.

The sculpting can be surprisingly precise. With code, one can choose exactly which antigen variant to express, stabilize a protein in a particular conformation, or include multiple antigens in one formulation. One can also tune where the antigen goes inside the cell by adding targeting sequences, influencing whether it is secreted, membrane-bound, or processed in particular compartments. Each choice changes how the immune system perceives the antigen: as an internal sign of infection, a surface marker, a soluble signal, or some combination. Vaccinology begins to look less like delivering a single object and more like designing a short-lived physiological state.

It also forces a clearer view of what immune memory "wants." The immune system is not a passive library in which you shelve an antigen and hope to find it later. It is an active economy that invests in certain memories and divests from others. To induce memory, a vaccine must create a situation in which it is rational, by the immune system's own internal logic, to invest: activate dendritic cells, provide co-stimulation, recruit helper T cells, and sustain antigen long enough for germinal centers to do their slow, painstaking work of improving antibodies.

Genetic instruction sets can create that situation by approximating one key feature of infection: the body's own cells producing a foreign protein, briefly, in a context that includes innate alarm. That is why they can be so effective even when they contain no whole pathogen and no traditional adjuvant in the older sense. The delivery vehicle and the intracellular event *are* the context.

The cowpox hack taught that resemblance can substitute for danger. Adjuvants taught that danger context can substitute for dirt. Genetic vaccines teach something even more audacious: instructions can substitute for matter. A few micrograms of nucleic acid—an amount too small to see without instruments—can provoke an immunological education that once required either surviving disease or risking it deliberately.

In the end, the most striking feature of the genetic instruction set is not technological but philosophical. It treats immunity as a programmable system, not because the immune system is simple, but because it is predictable in a particular way. Give it antigen in the right place, at the right tempo, with the right cues, and it will build memory. Not perfectly, not uniformly in every person, and not without trade-offs, but reliably enough that the world can plan around it.

A vaccine made of code does not feel like code inside the body. It feels, to the immune system, like an intrusion that half resembles infection: a cell behaving oddly, translating an unfamiliar script, displaying suspicious fragments on its surface. The immune system responds the way it has been shaped to respond—by learning, by remembering, and by preparing itself to decide faster the next time the pattern appears.

Chapter 8

Time and Tenants

Our immune system is not merely a genetic inheritance; it is a biography written in the ink of cellular memory. This chapter unravels how the trillions of microbes inhabiting our gut constantly tune our defenses, and how the relentless progression of time slowly transforms a nimble, naïve army into a rigid, experienced guard.

8.1 The Trillions Who Tutor Us

A newborn arrives with an immune system that is impressively equipped and, in one crucial sense, unfinished. The parts list is there: cells that can swallow intruders, cells that can make antibodies, cells that can kill infected targets, and cells that can coordinate everyone else. What is missing is calibration. An immune system that reacts to every molecule as if it were a tiger is not "strong"; it is exhausting, destructive, and often sick.

Calibration happens in conversation with the world, and nowhere is that conversation more constant than at the gut. The human intestine is a long, warm tube with an enormous surface area, folded and frilled to extract nutrients from food. It is also home to an astronomical number of microorganisms: bacteria, viruses that infect bacteria, fungi, and others. Popular descriptions call this community the "microbiome," and it is easy to picture it as a crowd of hitchhikers. A more accurate image is a tenant association: stable residents who pay rent, keep the hallways in order, and sometimes call the authorities when a stranger breaks in. The rent is chemical.

The immune system is often described in military language, but the intestine does not behave like a battlefield. It behaves more like a busy border crossing that must let through an endless stream of harmless travelers (food molecules) and long-term residents (microbes) while still catching the occasional smuggler (pathogens and parasites). The hard part is not attacking; the hard part is *not* attacking, and doing so without becoming blind.

The intestine's basic architecture makes the dilemma visible. A single layer of epithelial cells separates the body's interior from the intestinal contents. Those epithelial cells are not just tiles in a wall. They secrete mucus, produce antimicrobial peptides, and talk continuously to immune cells sitting immediately beneath them. If the barrier were fully sealed, the immune system would be ignorant of who lives outside. If it were wide open, inflammation would be constant. Evolution has settled on a compromise: carefully controlled sampling.

A classic and surprisingly vivid example comes from the way dendritic cells behave. Dendritic cells are professional "presenters": they take fragments of microbes and other material, process them, and show them to T cells as if holding up a photograph. In lymph nodes and gut-associated lymphoid tissue, T cells use these photographs to decide what kinds of responses to build. In the intestine, some dendritic cells extend thin projections between epithelial cells to reach into the gut lumen and sample what is there. It is a physical gesture of curiosity, like someone slipping a hand through a curtain to feel the weather before stepping outside. The point is not to invite invasion; it is to collect information that makes restraint possible.

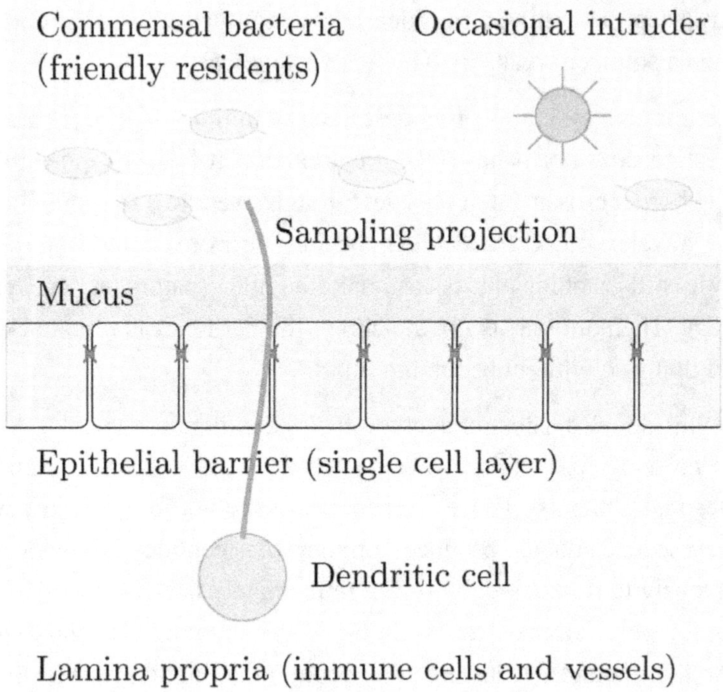

Sampling is only the first step. The deeper story is how the immune system turns those samples into a day-to-day "tone": a set of thresholds for responding. Too low a threshold, and harmless stimuli produce inflammation. Too high, and dangerous ones slip through. The microbiome participates in setting these thresholds through three broad routes: by occupying space and resources, by shaping the gut barrier, and by producing molecules that act as immune signals.

Consider the simplest effect: a crowded apartment building is hard to burglarize. Many commensal bacteria compete effectively for nutrients and attachment sites, leaving less room for invaders. Some produce bacteriocins, targeted toxins that suppress competitors. That is ecology, not immunity, but it has immune consequences. When the resident community is disrupted, infections become easier. The clearest real-world example is *Clostridioides difficile*. After a course of broad-spectrum antibiotics, the normal gut community can be thinned out, and *C. difficile* can seize the opportunity to expand, releasing toxins that inflame and damage the colon. In many patients with recurrent disease, replacing the missing community with a fecal microbiota transplant can be dramatically effective. The lesson is not romantic (no one should idealize "natural" microbes as automatically good), but it is concrete: the immune system's job is easier when the microbial neighborhood is stable.

The second route is the barrier itself. The gut lining must remain tight enough to prevent wholesale leakage of microbial products into the bloodstream, while still absorbing nutrients. Commensal microbes influence the production of mucus and antimicrobial peptides, and they encourage epithelial cells to maintain healthy junctions. When this partnership is stressed—by infection, inflammation, certain drugs, malnutrition, or severe disruptions in the microbiome—the barrier can become more permeable. Popular culture has turned "leaky gut" into a catch-all explanation for vague symptoms; the real biology is more precise and less sweeping. Increased intestinal permeability can occur, it can matter, and it can feed inflammation, but it is not a single diagnosis and not a universal cause. What it does illustrate is the everyday labor of a well-tuned gut: keeping the immune system informed without keeping it perpetually alarmed.

The third route is the most intimate: metabolites. Microbes are chemists. They take in components of our diet that we cannot digest fully—especially complex fibers—and transform them into small molecules that diffuse through the gut environment and into our tissues. Among the best studied are the short-chain fatty acids (SCFAs): acetate, propionate, and butyrate. These are not exotic compounds.

They are simple, and their simplicity is part of their power: they are abundant, fast to produce, and able to act on many cell types.

Butyrate is a particularly instructive example because it links food, microbes, and immune tolerance with a chain of cause and effect that is unusually clear. Many gut bacteria produce butyrate when they ferment dietary fiber. Colon epithelial cells use butyrate as a fuel, which helps maintain the barrier. Butyrate also influences immune cells, including T cells, by altering gene regulation. One mechanism involves inhibiting enzymes called histone deacetylases (HDACs). Histones are proteins around which DNA is wrapped, and chemical modifications to histones can make genes more or less accessible. HDAC inhibition tends to increase histone acetylation, which often makes certain genes easier to express. In the colon, butyrate can support the development of regulatory T cells (often abbreviated Tregs), a subset of T cells that dampens immune reactions and helps enforce tolerance. Tregs are not "anti-immune"; they are the immune system's braking system, essential for preventing collateral damage.

This is not merely a soothing story about calmness. The same class of microbial metabolites can, under different circumstances, support more aggressive T-cell programs. The gut must defend against pathogens that thrive in the intestine, and it must do so without setting the whole lining on fire. Another T-cell subset, Th17 cells, specializes in recruiting neutrophils and strengthening defenses against certain bacteria and fungi at barrier surfaces. Certain commensal organisms can promote Th17 development, which can be beneficial in infection. Yet Th17-skewing can become a liability if it pushes the immune system toward autoimmunity or chronic inflammation. In mouse models, particular gut microbes can increase susceptibility to autoimmune-like diseases by promoting pro-inflammatory T-cell responses. The point is not that a "bad microbe" causes autoimmunity in any simple way; rather, the immune system learns a style of responding in the presence of its microbial teachers, and that style can be helpful or harmful depending on context.

An anecdote from the history of laboratory animals shows how decisive these microbial teachers can be. In the mid-twentieth century, researchers began developing "germ-free" animals—especially mice—raised in sterile isolators, delivered by cesarean section, and kept away from all microorganisms. The original motivation was control: if you can remove microbes, you can reduce experimental noise and identify causes more cleanly. What emerged was not merely a cleaner mouse, but an odd one. Germ-free mice have underdeveloped gut-associated lymphoid tissues, altered numbers and types of immune cells, and a strikingly different immune baseline. Their immune systems are not just quieter; they are missing key forms of training. When scientists colonize germ-free mice with particular microbes or microbial communities, aspects of immune development can be restored, often in very specific ways. It was an early demonstration that immunity is not only genetically encoded. It is also ecologically educated.

This education begins early. Birth, feeding, antibiotics, infections, and environment all shape which organisms colonize the gut. The immune system, in turn, shapes the community by secreting IgA antibodies into the gut lumen. IgA is a specialized antibody class suited to mucosal surfaces. Rather than "killing" microbes in the way antibodies are sometimes imagined to do in the bloodstream, IgA often works by coating organisms and limiting how closely they can approach the epithelium, or by nudging microbial behavior. IgA can be thought of as a set of rules posted in the lobby: you may live here, but you do not get to enter every room. The rules are not the same for every resident. Over time, the immune system produces IgA that is targeted toward particular microbes and even toward dietary antigens. Food, too, is part of the tenant list. The gut constantly receives proteins from plants and animals that are foreign in the literal sense (not "self"), yet must be tolerated. The immune system learns that foreignness is not the same as danger.

This is where the immune system's decision-making becomes easiest to misunderstand. People sometimes speak as if the immune system "recognizes" friendly bacteria and therefore leaves them alone. In reality, recognition is not equivalent to permission. Immune cells can recognize commensal microbes very well; the gut is full of pattern-recognition receptors that detect microbial motifs. What distinguishes peace from war is the *context* in which recognition occurs: where the microbe is located, whether it is breaching the barrier, which danger signals are present, and which regulatory cues are abundant. A commensal bacterium that stays in the lumen, bathed in mucus and IgA, provokes a different response than the same bacterium injected into the bloodstream. The immune system is not contradicting itself in those two situations; it is applying a location-based rule that makes a barrier organ survivable.

A useful way to picture this is to imagine that immune responses are built from a limited set of "verbs": recruit, kill, repair, remember, restrain. The microbiome influences which verbs are most readily available on a typical day. A gut enriched in fiber-fermenting microbes producing plenty of SCFAs may bias toward barrier maintenance and tolerance, partly by supporting regulatory pathways. A gut destabilized by antibiotics, low-fiber diet, infection, or chronic stress can shift that balance, changing epithelial signals and immune cell programming. None of this is destiny, and none of it makes a simple morality play of "good" and "bad" microbes. It does mean that the immune system's baseline is not set by the body alone.

The popular appetite for probiotics reflects a real intuition: if microbes educate the immune system, perhaps we can choose our teachers. The difficulty is that classrooms are not interchangeable. A bacterium that survives well in one person's gut may fail to establish in another. Effects depend on diet, existing microbes, host genetics, and timing. A single strain sold in a capsule rarely recreates the stability and metabolic output of a complex community. That does not make all probiotics useless; it makes the promise easy to oversell. In many cases, the most reliable way to change microbial metabolism is not to add a microbe but to change the food supply. Fiber is not a vitamin, but it is a substrate for an ecosystem, and ecosystems respond more predictably to resources than to introductions.

There is also a cautionary tale embedded in the language of "boosting" immunity. A hyperreactive immune system is not necessarily desirable. Allergic disease offers a familiar example. The immune system treats pollen, peanuts, or pet dander as if they were parasites, launching responses that can be unpleasant or even life-threatening. Over the last several decades, the prevalence of allergic and autoimmune conditions has risen in many industrialized countries, inspiring hypotheses about modern environments. In 1989, the epidemiologist David Strachan proposed what became known as the hygiene hypothesis: smaller family size and cleaner early-life environments might reduce childhood infections and inadvertently increase allergy risk. The idea has evolved since then, and it is not a plea for dirt. The more refined view is that early-life exposure to a wider range of microbes—especially those associated with other humans, animals, and natural environments—can help the immune system establish robust regulatory pathways. The microbiome is one route through which that exposure may operate, though not the only one. The underlying theme is calibration again: a developing immune system needs examples of "harmless foreign" and "contained foreign" to avoid learning that every unfamiliar molecule merits panic.

Even in adulthood, the tutoring continues. The gut's immune system makes decisions on a short leash. Inflammation can damage the barrier and invite deeper invasion; too little responsiveness allows pathogens to gain a foothold. Commensal microbes constantly provide low-level stimuli that keep certain defenses primed without forcing a full-scale response. This is not the same as chronic inflammation. It is more like keeping a city's services running: streetlights on, water pressure steady, emergency services staffed. In fact, the gut's immune environment is rich in signals that actively promote restraint, including cytokines such as IL-10 and TGF-β, and a large population of regulatory immune cells. The presence of microbes helps justify that restraint: it gives the immune system a reason to specialize in discrimination rather than in indiscriminate force.

One of the most profound implications of this arrangement is that immune "memory" at barrier surfaces is not only memory of past infections. It includes memory of the body's long-term relationships. The immune system remembers, in a functional sense, what it is like to live with certain microbial communities: which responses are costly, which repairs are urgent, and which alarms are false. Some of that memory is held in classical adaptive immune cells—T cells and B cells shaped by repeated exposure. Some of it is held in the innate system's long-term tuning: altered responsiveness of macrophages and other cells after repeated stimulation, and durable changes in the barrier environment itself.

A quiet but consequential detail is that the immune system does not merely tolerate commensals in the abstract; it often tolerates them while still containing them. Many commensal bacteria would cause severe disease if they escaped the gut. The immune system therefore practices a form of measured suspicion: permissive at the surface, intolerant at depth. When this spatial discipline breaks down, the results can look like betrayal: inflammation directed against microbes that used to be neighbors, and tissue damage that further erodes the barrier. The tragedy is that the immune system is trying to restore order. It is simply doing so with fewer good options.

The gut, then, is not a sealed container but a managed interface. Food arrives, microbes flourish, and the immune system listens, samples, and decides. From this steady exchange comes a surprising conclusion: many of the most important immune decisions are not made in the drama of an acute infection, but in the routine of an ordinary day—in the choice to ignore a meal, to coat a bacterium with IgA rather than attack it, to instruct a T cell to become a regulator rather than a fighter, to repair a barrier rather than scorch it. The immune system learns what kind of world it lives in by meeting its tenants, and it repays the lesson by making coexistence biologically possible.

8.2 The Twilight of the Naïve

A child's immune system behaves like an eager linguist. It can learn new pathogens quickly, improvise responses, and store durable memories. By late adulthood, that improvisational talent is often harder to summon. Colds linger. Vaccines protect, but not as reliably. Shingles appears as a grim reminder that a virus can hide for decades and then resurface when cellular surveillance slackens. This gradual shift is called *immunosenescence*: the age-associated remodeling of immunity. The word can mislead, because it suggests a simple decline. What actually happens is closer to a change in staffing and budgeting. Some departments shrink, some expand, and the whole organization becomes more experienced in familiar problems but less adaptable to new ones.

The most consequential staffing change begins in an organ most adults rarely think about: the thymus. It sits above the heart, behind the breastbone, and in early life it is a thriving nursery for T cells. T cells are adaptive immune cells that can recognize specific molecular features (antigens) and either coordinate immune responses (helper T cells) or kill infected cells (killer, or cytotoxic, T cells). The thymus performs a delicate task: it generates a vast variety of T-cell receptors through genetic shuffling, then tests those receptors. Cells that cannot recognize the body's own molecular display at all are discarded; cells that recognize the body too strongly are also discarded. What remains is a repertoire that is both usable and restrained.

A historical episode made the thymus impossible to ignore. In the early 1960s, Jacques Miller performed experiments in which newborn mice were thymectomized—their thymus removed soon after birth. The result was not subtle. The mice developed profound immune defects, including an inability to mount normal immune responses. In an era when the thymus was sometimes dismissed as a vestigial organ, these experiments helped establish it as central to T-cell development. That discovery still echoes in modern medicine: therapies that damage the thymus or the bone marrow (such as chemotherapy) can leave patients immunologically fragile, and the speed and completeness of immune reconstitution often depend on how much thymic function remains.

The thymus does not stay youthful. Beginning relatively early in life, thymic tissue gradually involutes: functional thymic epithelium is replaced by fat, the architecture becomes less supportive, and the output of new T cells declines. This matters because newly produced T cells are, by definition, *naïve*. "Naïve" does not mean foolish; it means uncommitted. A naïve T cell has not yet encountered its specific antigen. It carries a unique receptor, representing a lottery ticket against future infections. A large pool of naïve T cells gives a person breadth: the ability to respond to unfamiliar microbes because, somewhere in that diverse pool, a few cells will recognize the new threat and expand.

With age, the balance in the T-cell population shifts. Memory T cells accumulate. These are cells that have seen antigens before and have been shaped by those encounters. Memory is invaluable: it is why a second encounter with measles, or a successful vaccine, can produce rapid protection. But memory has a trade-off. Memory cells occupy space, consume resources, and compete for survival signals. Meanwhile, fewer new naïve cells arrive from the thymus. The immune system becomes top-heavy with experience.

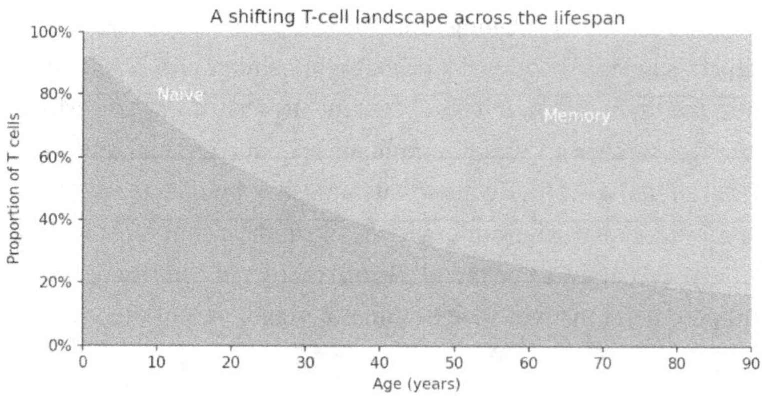

The figure is intentionally simple: real people vary enormously, and different tissues (blood, lymph nodes, mucosal sites) have different mixtures. Still, the overall direction is robust. Early life is dominated by naïve cells; later life is dominated by memory and memory-like cells. The mechanical consequence is straightforward: fewer fresh recruits means fewer chances to recognize a brand-new pathogen quickly.

Yet the immune system is not passive in the face of thymic decline. It compensates. Naïve T cells can persist for years, and they can divide slowly in the periphery to maintain their numbers. This process is called *homeostatic proliferation*. The name sounds calm, but it carries a subtle cost. When a cell divides, it is no longer exactly the same cell; division selects for cells that respond better to the survival signals available. Over many years, this can bias the naïve pool toward certain clones and certain behaviors. The pool may look numerically adequate while becoming less diverse in practice.

Immunologists have devised clever ways to infer how much thymic output remains. One approach uses T-cell receptor excision circles (TRECs), small circular DNA byproducts created during receptor gene rearrangement in the thymus. TRECs do not replicate when a cell divides, so they become diluted with proliferation. High TREC content suggests a cell is a "recent thymic emigrant," newly arrived from the thymus; lower TREC content suggests a longer peripheral history. Another approach in humans uses surface markers such as CD31 on naïve CD4 T cells to distinguish subpopulations enriched for more recent thymic emigrants. None of these measures is perfect, but they reveal the same broad narrative: thymic contribution wanes, and peripheral maintenance becomes a bigger part of the story.

Aging also reshapes the immune system's internal geography. The lymph nodes and spleen are not just bags of cells; they are carefully organized meeting places where dendritic cells present antigens, T cells and B cells converse, and immune responses are choreographed. With age, the stromal scaffolding and the chemistry of these sites can change. The result is less efficient matchmaking between rare antigen-specific naïve cells and the signals they need to expand. Even if a relevant naïve clone exists, the odds of it being recruited promptly can fall.

Then there is the question of *quality*, not just quantity. Many aging T cells show signs of replicative wear. They may have shorter telomeres (protective chromosome ends that shorten with division), altered metabolic capacity, and different patterns of cytokine production. Some acquire features of "exhaustion" or "senescence-like" states, especially after chronic stimulation. These words are sometimes used loosely; it is safer to say that repeatedly activated T cells can enter long-lived states that prioritize immediate effector function or survival over flexibility and expansion.

Chronic infections can act like steady winds shaping the immune landscape. Cytomegalovirus (CMV), a common herpesvirus, is a famous example. Once acquired, it persists for life, usually without causing obvious illness in healthy adults. But it is not silent. In many people, CMV drives large expansions of particular memory CD8 T-cell clones—cells specialized for killing infected targets. These expanded populations can become a substantial fraction of the circulating T-cell pool in older adults. The image is dramatic: an immune system increasingly populated by veterans of a conflict that never entirely ends.

What, exactly, is the cost of these expansions? For years, a popular story held that CMV "crowds out" the repertoire, pushing aside diversity and leaving fewer slots for responses to new pathogens. More recent high-throughput studies complicate that narrative. When researchers account properly for the dominance of a few expanded clones, underlying repertoire diversity may not be as catastrophically reduced as the crowding-out metaphor implies. The immune system can be spacious in ways our metaphors miss. Still, even if diversity remains, the *distribution* of attention changes. A pool in which many resources are tied up in long-lived effector-memory populations may respond differently to a novel vaccine antigen than a pool with abundant naïve reserves.

Vaccination offers a concrete window into these changes, because it is an attempt to create immune memory on demand. In older adults, many vaccines produce lower antibody titers, less durable protection, or weaker cellular responses. That is not because older immune systems "cannot learn." They can and do. It is because the process of learning relies on a chain of events: antigen presentation, T-cell help, germinal center formation in lymphoid tissues, B-cell selection, and the generation of long-lived plasma cells and memory B cells. Aging nudges many links in that chain toward lower efficiency. This is why some vaccines are formulated differently for older adults, using higher antigen doses or adjuvants (immune-stimulating additives) that provide stronger "danger" context and improve recruitment of the cells that remain capable of robust expansion.

The thymus also contributes to another aspect of immune aging that is easy to overlook until it becomes personal: tolerance. Central tolerance, the process of deleting strongly self-reactive T cells in the thymus, depends on the thymus doing its job well. As thymic structure and function change, the details of negative selection may change too. Autoimmunity in older age is not simply a matter of "the immune system attacking the self because it is confused," but altered tolerance is one plausible ingredient. Another is that tissues accumulate damage with age. Damaged cells release molecules that act as danger signals; proteins become modified; senescent cells secrete inflammatory factors. The immune system is asked, more frequently, to interpret ambiguous signals coming from stressed tissues. The boundary between repair and attack becomes harder to police.

This leads into *inflammaging*, the chronic, low-grade inflammation associated with aging. It is tempting to blame inflammaging on an "overactive immune system," but the deeper pattern is dysregulation: a persistent background of inflammatory signals alongside weaker responses to new infections. How can both be true? One answer is that inflammation is not always a sign of effective defense. Inflammation is a tool. If the tool is used imprecisely, you can get more sparks and less work.

Some of the inflammatory background in older adults may come from innate immune cells that have been repeatedly stimulated over a lifetime and have adopted different set-points. Some may come from the accumulation of senescent cells and debris that the immune system struggles to clear efficiently. Some may come from barrier changes in the gut and lungs that allow more microbial products to leak into the body's interior, providing constant mild provocation. None of these explanations needs to be universal to be meaningful. Together, they point to a system that is spending more energy on housekeeping fires and has less reserve for sudden emergencies.

Despite all this, it would be a mistake to imagine immune aging as a one-way slide into incompetence. The experienced immune system has strengths. Memory can be extraordinarily protective. An older person who has been vaccinated and exposed to seasonal viruses for decades often has a layered collection of antibodies and T-cell memories that provide partial protection even when a pathogen changes. Sometimes the immune system's problem is not that it cannot respond, but that it responds in a way that reflects its history. A response optimized for old threats may be deployed against a new one, with mixed results.

The possibility that immune aging is modifiable is no longer merely wishful thinking. Thymic involution, once treated as an inevitability, appears to have some plasticity. Researchers have explored interventions ranging from cytokines such as IL-7 (which supports T-cell survival and development) to growth factors and hormonal manipulations that can partially rejuvenate thymic function in certain contexts. None of these is a simple prescription for healthy aging, and any attempt to "rejuvenate" immunity must contend with the risk of promoting autoimmunity or malignancy. Still, the very fact that thymic function can move at all is revealing. The immune system is not a clock that runs down; it is a living tissue responsive to signals, stress, infection, nutrition, and repair.

A more grounded way to think about immunosenescence is as a shifting portfolio. In youth, the immune system invests heavily in options: a wide range of naïve receptors, many of which will never be used, but which provide readiness for surprises. With time, the portfolio tilts toward assets with a proven track record: memory cells expanded by real encounters, sometimes maintained for decades. The return on that investment is rapid response to familiar threats. The risk is that novelty becomes expensive. When a new pathogen arrives, the immune system must search a smaller, more uneven pool for rare matching naïve clones, recruit them through lymphoid tissues that may be less efficient, and expand them in a body already managing a higher inflammatory baseline.

One can see the human cost of this shift most clearly in diseases that punish delay. Influenza has long been more dangerous in the elderly, and the same pattern appears with other respiratory infections. The early days of a new infection are a race: if the immune system can contain viral replication quickly, tissues suffer less damage and the later inflammatory response can remain proportionate. If early containment is slower, the body may compensate with a broader, more damaging inflammatory response. The tragedy is that the later inflammation can look like strength when it is actually the smoke from a fire that was allowed to spread.

The twilight of the naïve is not the end of immune memory; it is, in a sense, the price of having accumulated so much of it. Every memory cell represents a past victory, or at least a past battle survived. The question is not whether the immune system should remember, but how it can continue to make room for the unknown while carrying the weight of what it has already learned. In the aging body, that balance is negotiated every day in quiet cellular arithmetic: which cells are allowed to persist, which are replaced, which signals are amplified, and which are restrained.

8.3 The Biography of Exposure

Two people can sit side by side on the same train, breathe the same air, and share the same headline about a new virus, yet their immune systems will not meet that moment as equals. One may have a quick fever and be well in two days; the other may cough for weeks. One may feel little after a vaccine; the other may be knocked flat for an afternoon. These differences tempt us to speak in the language of strength and weakness, as if immunity were a single muscle. A more faithful picture is that each immune system carries a private archive: a record of infections, vaccinations, diets, pollutants, stresses, and microbial companions, all layered into the tissues. The archive is not stored like a diary; it is stored as cells and circuits that have been altered by experience.

Even identical twins, whose genomes begin as close to indistinguishable as biology allows, diverge immunologically with time. That divergence is not a failure of genetics; it is the expected consequence of living. One twin catches mononucleosis at fourteen, the other does not. One develops asthma after a smoky apartment and a series of winter viruses, the other grows up with a dog and spends weekends outdoors. One takes repeated antibiotics for acne, the other never does. The immune system is built to remember, and it remembers whatever life insists on teaching it.

A useful term for this accumulated history is the *exposome*: the total of environmental exposures and interactions across a lifetime. The word is sometimes used so broadly that it becomes foggy. For immune memory, it becomes clearer if you narrow the question. Which exposures leave lasting marks on immune decisions? Which change the set-points for responding, the repertoire of remembered antigens, and the likelihood of collateral damage?

Some exposures are dramatic and easy to name. A measles infection can erase aspects of immune memory to other pathogens, leaving children temporarily more vulnerable afterward; a vaccine can prevent that catastrophe. A severe influenza infection can leave a scarred lung that is physiologically different for years. A course of chemotherapy can wipe out immune cell populations and force the body to rebuild. These events are like earthquakes in the immune landscape.

Most immune biography, however, is written in small daily entries. The gut microbiome, already established as a constant tutor, keeps adjusting what counts as normal background signal. The thymus, gradually reducing the supply of naïve T cells, shifts the system's ability to respond to the unfamiliar. Onto that foundation, the immune system layers specific memories: memory B cells and T cells trained on particular viral fragments, bacterial proteins, and vaccine antigens. It also layers less specific changes: enduring shifts in the behavior of innate immune cells that have been repeatedly stimulated.

To understand how a lifetime becomes biology, it helps to separate two kinds of remembering. The first is classic adaptive memory. B cells and T cells recognize specific antigens, expand when they meet them, and leave behind memory populations. If you were vaccinated against tetanus as a child, you still have memory cells and long-lived plasma cells that can respond rapidly if the toxin appears. In many cases, that memory is exquisitely specific: it depends on the precise shape of a molecular epitope, the portion of an antigen that immune receptors actually bind.

The second kind of remembering is sometimes called *trained immunity*. Innate immune cells—macrophages, monocytes, natural killer cells, and even their progenitors in the bone marrow—can enter longer-lived states of altered responsiveness after certain infections, vaccines, or inflammatory episodes. Unlike adaptive memory, trained immunity is not based on a rearranged receptor that recognizes a particular epitope. It is based on changes in gene regulation and metabolism: epigenetic marks that make certain inflammatory genes easier or harder to activate, and metabolic programs that change how readily a cell can mobilize energy for defense. Training can be helpful, providing broader readiness for unrelated infections. It can also be harmful, promoting chronic inflammation when the stimulus is gone.

These two modes of memory interact. A person's immune biography is not simply a list of antigens encountered. It is also a story of how readily their innate system raises alarms, how easily their tissues drift into inflammatory states, and how effectively regulatory mechanisms apply the brakes.

A concrete example of immune biography comes from influenza, a virus that has been circling humans for centuries and has left its fingerprints on generations. Influenza's surface proteins mutate, creating new strains that partially escape existing immunity. The immune system responds to influenza with both antibodies and T cells, and the details of that response depend strongly on the first influenza encountered in childhood. This phenomenon has been called *immune imprinting* or, in an older and more provocative phrase, *original antigenic sin*. The "sin" is not moral; it is architectural. Early exposures recruit certain B-cell clones that recognize particular features of the virus. When a related but different influenza strain appears years later, the immune system often returns to those earlier solutions, expanding memory cells that fit the old shape reasonably well rather than starting from scratch with naïve cells that might target new epitopes more effectively.

This is not always bad. If the new strain is similar enough, leaning on old memory can provide rapid partial protection and reduce severe disease. The problem arises when the similarities are misleading:

when the immune system's first draft is repeatedly revised rather than rewritten. The response can become biased toward conserved but non-neutralizing targets, or toward epitopes that are no longer the most important vulnerabilities. Researchers have debated how often imprinting harms vaccine effectiveness versus how often it helps by providing broad coverage. The answer seems to depend on the distance between strains and on which viral features are being targeted. What matters for immune biography is the principle: the immune system does not approach each exposure as an isolated event. It approaches the present through the lens of the past.

A historical anecdote makes the stakes of influenza history visceral. In 1918, an influenza pandemic swept the world with extraordinary lethality. One of its eerie features was its age pattern: young adults, who would normally have been expected to fare better, died in large numbers. Many factors were at play, including viral virulence and the possibility of immune overreaction. But the episode also reminds us that the immune system's past is not always protective. A strong immune response can be misdirected; the wrong kind of inflammatory program can damage lungs. "Experience" is not synonymous with wisdom, and sometimes the price of responsiveness is collateral injury.

Immune biography extends beyond viruses. Consider allergies. A child growing up in an environment with high exposure to certain allergens, with intermittent viral infections, and with air pollution that irritates the airway epithelium may be pushed toward a particular style of immune response: one that favors type 2 inflammation, involving eosinophils and IgE antibodies. Another child with different microbial exposures and fewer epithelial injuries may develop more robust regulatory networks and a different balance of helper T-cell programs. These are not simple cause-and-effect chains, and no single exposure guarantees asthma or protects against it. The point is that the immune system learns not only what to attack, but how to interpret irritation, damage, and repeated harmless encounters.

Pollutants deserve a place in the immune biography because they can alter the context in which antigens are perceived. Particulate matter from combustion, cigarette smoke, and certain industrial exposures can inflame and damage mucosal surfaces, changing barrier function and changing the signals immune cells receive. An antigen encountered in a calm tissue may be tolerated; the same antigen encountered during epithelial injury may be tagged as suspicious. Immunity is deeply sensitive to *where* and *when* an antigen appears.

Diet, too, leaves immune fingerprints. The gut is exposed to dietary antigens constantly, and the immune system must maintain tolerance while still producing protective IgA and responding to pathogens. Diet shapes the microbiome's metabolism, influencing metabolites such as short-chain fatty acids that affect regulatory T cells and barrier integrity. A long-term low-fiber diet can reduce the production of these metabolites and alter the microbial community, potentially nudging immune tone toward inflammation. Conversely, a diet that supports a stable, diverse microbial ecosystem may support mucosal tolerance. These are probabilistic influences, not guarantees, but over decades probabilities become biographies.

Vaccination is often treated as a clean intervention: an antigen delivered in a controlled dose, ideally producing a predictable memory. In reality, vaccination is also woven into immune history. A person's baseline inflammation, their microbiome, their age-related thymic output, and their prior encounters with similar antigens all influence vaccine responses. Sometimes the difference between a robust response and a weak one is not the vaccine itself but the immune context into which it lands. This has practical consequences. It helps explain why adjuvants can be especially useful in older adults: they provide stronger innate signals that recruit help for adaptive memory when baseline responsiveness is lower. It also helps explain why timing matters. A vaccine given during an acute infection, or during a period of high stress and poor sleep, can elicit a different response than the same vaccine given when the body is calmer.

The most unsettling aspect of immune biography is that it can lock in vulnerabilities. Chronic infections, for example, do not simply add one more memory; they can reshape the immune system's distribution of attention for decades. A persistent herpesvirus such as CMV can drive expansions of certain memory T-cell clones. Even if overall repertoire diversity is not catastrophically reduced, the immune system's resources are nonetheless allocated differently. In an older adult, a large fraction of the T-cell pool may be devoted to monitoring a virus that rarely causes symptoms, while fewer resources are available for responding to new threats. This is not a moral failing of the immune system; it is a consequence of living in a world where some adversaries never fully leave.

Immune biography also helps make sense of why aging is so heterogeneous. Chronological age is a crude proxy for immunological age. Two seventy-year-olds can have immune systems that behave more like those of fifty-year-olds or eighty-five-year-olds, depending on lifetime exposures and health. One person may have lived with chronic periodontal disease, obesity-associated inflammation, repeated respiratory infections, and long-term exposure to air pollution. Another may have had fewer infections, less chronic inflammation, more physical activity, and different microbial exposures. The difference is not merely lifestyle virtue; it is the cumulative burden of challenges and repairs that the immune system has managed.

Here, "inflammaging" becomes less mysterious. Chronic low-grade inflammation can be seen as the sum of many small signals: senescent cells that are not cleared efficiently, damaged tissues that leak distress molecules, metabolic stress, persistent infections, and perhaps increased microbial translocation across aging barriers. The immune system is responding to something real, but the response is no longer tightly coupled to a solvable target. The inflammatory program becomes background music. Background music changes how you hear everything else.

A striking feature of immune biography is its stubbornness. Memory cells can persist for decades. Epigenetic programs in innate cells can endure. The microbiome can be resilient, sometimes returning to its previous composition after perturbation, and sometimes flipping into a new stable state. This persistence is both blessing and curse. It is why childhood vaccination can protect a lifetime. It is also why early exposures can shape later responses in ways that are hard to undo.

And yet immune biography is not fate carved in stone. The immune system remains plastic because the world keeps changing. New exposures arrive; old infections are controlled; barriers heal; diets change; sleep improves; stress eases; a new vaccine teaches a new epitope. Plasticity has limits, especially as thymic output declines and tissue repair becomes slower, but the immune system continues to negotiate.

The idea of a personal immune fingerprint also changes how one should interpret tests and labels. Measuring antibody levels in blood provides a glimpse of one part of the archive, but not the whole. A person can have modest circulating antibodies yet retain strong memory B cells capable of rapid recall. Another can have detectable antibodies but poor T-cell help. Some immune memories live mainly in tissues, not in blood: resident memory T cells in the lungs or gut that respond quickly at the site of entry. Even the same exposure can leave different kinds of memory depending on route and context. An infection acquired through the nose may seed lung-resident memory; an injected vaccine may emphasize systemic antibodies. The immune fingerprint is multi-layered and location-specific.

If there is a unifying lesson in the biography of exposure, it is that the immune system's decisions are always made with incomplete information and a long memory. It must guess, based on patterns: how dangerous is this signal, how often has it accompanied harm, how costly is it to respond, how costly is it to wait. In youth, the system can afford to be exploratory, supported by a generous supply of naïve cells and a relatively clean inflammatory background. With time, the system becomes more opinionated. It responds faster to familiar cues and can be slower or more awkward with novelty.

At the scale of a life, immune memory looks less like a library of single titles and more like handwriting: recognizable, personal, shaped by early lessons and later habits. Some people's immune handwriting is spare and calm, with strong tolerance and efficient recall. Others carry heavier strokes: more inflammation, more scarring, more reactive defaults. No two immune systems live in the same environment, even when they share an address, because the environment that matters includes the past.

A clinician listening to a patient's story often senses this without using the term "immunobiography." The patient who had severe childhood asthma, repeated steroid courses, and frequent infections may respond differently to a respiratory virus in adulthood. The patient who worked for decades in a polluted industrial setting may have an airway epithelium primed for irritation. The patient with a history of CMV or other persistent infections may have an immune profile skewed toward certain memory states. These are not excuses; they are explanations. They move immunity away from simplistic judgments and toward a more humane realism: bodies are shaped by what they have endured.

The immune system cannot choose its first lessons, and it cannot erase them cleanly. It can, however, keep learning. In that ongoing education lies both vulnerability and hope: vulnerability because the wrong lesson can echo for years, hope because new lessons can still be written into living tissue, molecule by molecule, cell by cell, until a once-unfamiliar threat becomes just another remembered shape.

Chapter 9

Decoding the Black Box

For decades, immunologists described the immune system by looking at the average behavior of millions of cells, effectively listening to a crowd roar rather than individual conversations. We now possess the tools to dismantle this biological black box, measuring the precise genetic state of single cells and mapping the mathematical logic that governs their decisions. This quantitative revolution transforms immunity from a descriptive biological art into a predictable engineering science.

9.1 Eavesdropping on the Cellular Monologue

Imagine being told you may listen to a stadium crowd, but only through a single microphone hung high above the field. You would hear a roar that rises and falls, and you might even pick out a chant if it is loud enough. Yet the crowd is not a single voice. It is thousands of people: some singing, some arguing, some whispering, some leaving early, some sitting in stunned silence. For a long time, much of immunology was that overhead microphone. Researchers took a piece of tissue—a lymph node, spleen, blood sample, tumor biopsy—and measured the average activity of all the cells inside. The results were immensely valuable. They told us which genes were "on" during infection, which proteins rose after vaccination, which inflammatory pathways surged in autoimmunity. But the method smoothed away the individuality that makes the immune system work.

Immune memory, in particular, thrives on individuality. Memory cells are not a uniform reserve army waiting in identical tents; they are a distributed, opinionated society. Some stand guard in tissues, some patrol the bloodstream, some specialize in calling for help, and some are shaped by past battles in ways that only show up when the right challenge appears. The difficulty is that the body does not label these roles with tidy tags. It leaves clues scattered across thousands of molecular details inside each cell. Single-cell sequencing is the technology that finally let scientists hear the crowd as a crowd.

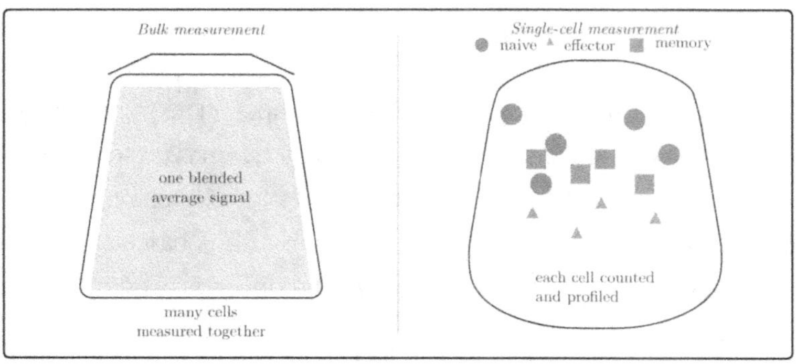

The old approach is usually called *bulk* measurement. If you grind up a million immune cells and measure RNA—the working copies of genes that a cell is actively using—you get a single profile: the average transcriptome of the mixture. The average can be informative. It can also be deceptive. If half the cells are highly activated and half are quiet, the average looks like a moderately activated cell that does not actually exist. If a rare but crucial cell type makes up one in a thousand cells, its signature may vanish into the background noise. For immune memory, rarity matters: the cells that prevent reinfection can be a small minority, and their protective role may depend on being in the right place with the right readiness.

The most intuitive leap of single-cell sequencing is simply this: instead of measuring the smoothie, measure each piece of fruit. In practice, that means taking a suspension of cells and isolating them one by one, so that the RNA from each cell is captured, labelled, and sequenced separately. When the sequencing data come back, each cell has its own gene-expression "speech." The result is less like a photograph and more like eavesdropping on thousands of simultaneous monologues.

The story behind this leap has a characteristically modern twist: it is not only a biological discovery, but an engineering trick. A single immune cell contains a tiny amount of RNA. If you try to sequence it directly, you risk losing most of the message. Early single-cell methods were painstaking and low-throughput. The transformation came when researchers learned how to attach *barcodes* to molecules so that, after everything is pooled together for sequencing, you can still tell which molecules came from which cell. A barcode is a short stretch of DNA used as an ID tag. Add a different barcode to each cell, mix all cells into one tube, sequence everything, and then use the barcodes to sort the reads back into per-cell profiles. The trick is to do it reliably at scale.

Droplet-based methods made that scale practical. They turn a cell suspension into a stream of microscopic droplets, each ideally containing a single cell and a bead coated with DNA primers. The bead carries a cell-specific barcode and often a *unique molecular identifier* (UMI), another short tag that marks individual RNA molecules so that a heavily amplified transcript does not masquerade as many original transcripts. The cell is lysed inside the droplet; its RNA binds to the primers; reverse transcription converts RNA to DNA; and the barcode becomes physically attached to every captured molecule. Later, in the sequencing data, that barcode is the cell's name.

Once you have per-cell gene expression, the mind immediately asks: how many different immune cell types are there? The old textbook categories—"naive," "effector," "memory"—sound like clearly separated species. Single-cell data revealed something messier and more interesting: immune cells are often better understood as *states* rather than fixed types. A T cell can slide along a continuum of activation, stress, migration readiness, metabolic mode, and cytokine production. It can occupy transitional states that are brief but decisive. It can carry lingering molecular scars from past activation that do not show up in its surface markers. Two cells with the same name on a flow cytometry plot can behave differently because their internal programs differ.

A concrete example comes from influenza, a virus that has served as a laboratory training partner for immunology for decades. After flu infection or vaccination, some T cells become short-lived fighters, producing potent antiviral molecules and then dying off. Others become memory cells that persist. Even within memory, there are several flavours. Some circulate broadly through blood and lymph; others settle in tissues like the lung and remain there, a local militia. Before single-cell sequencing, separating these groups relied on surface proteins used as markers. Those markers are useful, but they are a partial vocabulary. Single-cell RNA sequencing added paragraphs. It revealed, for example, that cells classified as the same memory subset can carry different transcriptional programs: some primed for rapid cytokine release, some primed for proliferation, some primed for tissue residency, some quietly expressing stress-response genes as if still remembering the battlefield.

The most surprising discoveries often come not from confirming expected categories, but from seeing unexpected minorities. In a bulk sample, a rare cluster of cells is just a rounding error. In single-cell data, it can stand out as a distinct island of expression. Sometimes that island corresponds to a known but elusive population; sometimes it is something new. This is especially relevant in human immunology, where researchers cannot ethically do many of the controlled infections and tissue sampling that are routine in animal models. Human immune memory must be inferred from what can be sampled—usually blood—and from the imperfect natural experiments of infection and vaccination. If the decisive memory population sits in the nasal mucosa or deep in the lung, blood can be a misleading window. Single-cell methods have at least made that window sharper: within blood, they can find rare activated cells, rare antigen-specific cells, and rare cells with tissue-homing signatures.

This is where a brief historical anecdote helps explain why the field was hungry for such sharpness. In the 1970s, César Milstein and Georges Köhler developed the hybridoma technique to make *monoclonal antibodies*: antibodies produced by a single B cell clone, each recognising one specific target. That breakthrough gave immunology its most precise molecular probes, and it changed medicine by enabling targeted therapies and diagnostics. But it also reinforced a certain kind of thinking: if you can isolate one clone and learn its specificity, perhaps the immune system is best understood clone by clone, in discrete categories. The decades that followed showed that specificity is only part of the story. Two B cells with identical antibody genes can behave differently depending on where they are, what signals they received, and what epigenetic and transcriptional programs they carry. Single-cell sequencing, in a sense, took the monoclonal mindset—*one cell at a time*—and made it scalable enough to reveal the social structure of immunity rather than just its individual characters.

Single-cell sequencing does more than catalogue diversity; it forces a rethinking of what "identical" means. Consider two memory T cells that recognise the same viral epitope (the small part of an antigen that a receptor binds). Traditional assays might group them as the same clone or the same subset. Yet their RNA profiles can suggest different histories. One may express genes associated with tissue migration and adhesion, as if it has learned to take up residence. Another may express genes that keep it poised to divide rapidly, as if its role is to expand a response when reinfection occurs. Another may show a signature of recent interferon exposure, a hint that it has been patrolling in a virus-rich environment. These differences are not decorative. They are how the immune system hedges its bets: it spreads risk across strategies.

A common worry, when first hearing about single-cell RNA sequencing, is whether RNA truly reflects function. RNA is a plan, not a finished product. Proteins are the working parts of the cell: receptors, enzymes, cytokines, transcription factors. RNA levels often correlate with protein levels, but not always. Some proteins are stable and linger long after their mRNA is gone; some are rapidly degraded; some are regulated by translation rather than transcription. Immune cells also use surface proteins as badges that matter directly for their interactions. If you measure only RNA, you can misclassify cells that share a transcriptome but differ in surface phenotype, or miss clinically relevant distinctions that are protein-based.

This led to a clever hybrid approach: simultaneously measuring RNA and selected proteins in the same cell. CITE-seq, developed by Marlon Stoeckius and colleagues (2017), uses antibodies tagged not with fluorescent dyes, as in flow cytometry, but with short DNA oligonucleotides. When the cell is processed for sequencing, those oligonucleotides are captured and counted along with RNA, giving a quantitative readout of surface proteins in the same digital framework. The result feels like giving each cell both an internal diary

(RNA) and a set of visible clothes (surface markers) and reading both at once. In immune memory studies, this matters because many of the classic distinctions between naive, effector, and memory cells were built on surface markers, while functional specialisation often shows up in transcriptional programs. Combining the two reduces ambiguity.

There is another problem that only becomes obvious when you try to listen to thousands of cellular monologues: sometimes you are accidentally listening to two cells at once. Droplet methods aim for one cell per droplet, but biology is messy and statistics are indifferent. Occasionally two cells land in the same droplet, producing a mixed profile that looks like a strange hybrid state. If you are hunting for rare transitional populations, doublets can imitate them. In the worst case, a technical artefact becomes a published "new cell type."

To mitigate this, researchers developed strategies such as *cell hashing* (Stoeckius and colleagues, 2018). The idea is to label cells from different samples with different DNA-barcoded antibodies before pooling them. After sequencing, the hash tags reveal which sample each cell came from. If a droplet contains two cells from different samples, it will carry two hash tags and can be flagged as a doublet. Multiplexing also reduces batch effects: instead of processing samples on different days and worrying that day-to-day variation will masquerade as biology, you can pool them and process them together. This kind of methodological humility is part of what makes single-cell science persuasive. The technology is powerful enough to seduce, so it must also be self-critical.

Single-cell data also introduced a new kind of uncertainty: interpretation. When you turn thousands of cells into a cloud of points on a plot (often a two-dimensional map made by algorithms such as t-SNE or UMAP), clusters appear. It is tempting to name each cluster as a cell type, as if nature has kindly arranged discrete drawers. Sometimes it has. Often it has not. The immune system loves continua: activation gradients, differentiation trajectories, stress responses, cell-cycle phases. Clusters can reflect technical effects (batch, doublets, ambient RNA), or they can reflect biological processes that cut across types (interferon response, heat shock, hypoxia). A good single-cell study is therefore as much about scepticism as discovery. It uses multiple lines of evidence: RNA, protein, receptor sequences, and knowledge of immunology's logic.

One of the most elegant additions, especially for adaptive immunity, is measuring the immune receptors themselves. T cells and B cells carry receptors generated by gene rearrangement, producing an enormous diversity of specificities. With suitable methods, single-cell sequencing can capture the sequences of T-cell receptors (TCRs) or B-cell receptors (BCRs) alongside the transcriptome. This connects three things that used to be hard to connect at once: what a cell is (its gene-expression state), what it recognises (its receptor), and how it might behave (its functional program). A memory response can then be viewed as a set of clones distributed across states: some members of a clone are tissue-homing, some proliferative, some quiescent. Clonal identity becomes a thread running through a tapestry of fates.

A concrete, human-scale example is vaccination. After a vaccine, blood contains a transient wave of activated cells: plasmablasts producing antibodies, T cells proliferating, innate cells producing cytokines. In bulk data, this wave is a blur. In single-cell data, it becomes a timeline of characters entering and leaving the stage. Researchers can see, for instance, that a small number of antigen-specific T cell clones expand and that within those clones, some cells express genes associated with immediate effector function while others express genes associated with memory formation. This heterogeneity is not just noise; it is a potential explanation for why two people with similar antibody titres can have different durability of protection. One person's response may have quietly invested more in long-lived memory programs, while another's response spent its resources on short-lived effector activity. These are the sorts of differences that matter for public health and for personal risk, and they were largely invisible when we listened only to the average.

Single-cell approaches have also been revealing in contexts where immune memory is harmful, such as allergy and autoimmunity. In these diseases, the immune system "remembers" the wrong target or responds too strongly to a harmless one. Bulk measurements can show inflammation; they struggle to reveal which specific populations are sustaining it. Single-cell data can isolate rare autoreactive-like phenotypes, identify exhausted or regulatory programs that fail to restrain disease, and distinguish infiltrating immune cells from tissue cells responding to damage. The hope is not merely to catalogue, but to point to interventions that are more precise than blanket immunosuppression: therapies that remove or reprogram particular cell states while preserving the rest of immune competence.

All of this might suggest that single-cell sequencing gives a perfect, literal transcript of cellular thought. It does not. Cells are not minds, and their molecular states are not conscious narratives. They are chemical and genetic programs responding to signals. The monologue metaphor is useful because it captures two truths: each cell has an internal state that matters, and that state can be read as information. But like any eavesdropping, it is partial and context-dependent. The act of isolating cells from tissue can stress them; the captured RNA is a snapshot; some molecules are missed; the sequencing depth is finite. What the technology offers is not omniscience, but a new resolution at which the immune system becomes legible.

That new resolution has changed the emotional tone of immunology. The immune system used to look like a set of named troops marching in formation. Now it looks more like a crowded city at night, with lights flickering on and off in different windows, patterns emerging only when you watch many buildings at once. Immune memory, seen this way, is not a single stored object but a collection of prepared stances distributed across cells: readiness, restraint, mobility, endurance, and a catalogue of prior encounters written into receptor sequences and regulatory programs. Single-cell sequencing did not invent that complexity. It merely gave us the microphones sensitive enough to hear it.

9.2 The Signature of Safety

A child receives a vaccine on a Tuesday, develops a sore arm by evening, and by Friday is back at school. What, exactly, has been achieved? The most honest answer is probabilistic. The child is *less likely* to get sick, less likely to be hospitalized, less likely to spread infection. Yet medicine and public health often need a firmer-sounding statement: safe or unsafe, protected or unprotected. The immune system does not speak in binaries, and it does not leave a simple stamp of approval. It leaves patterns.

The practical problem is obvious as soon as it is stated. The clearest way to test immunity is to expose someone to a pathogen and see whether they get ill. For many diseases that is unethical, sometimes impossible, and often too slow. Even when challenge studies are possible, they use carefully selected volunteers and controlled doses that do not capture the full messiness of real life. Waiting for large numbers of infections to accumulate after a vaccine trial is costly and, in an epidemic, morally uncomfortable. If a biological measurement could stand in for the outcome—a laboratory value that reliably predicts protection—it would change everything: faster vaccine development, faster updates when a virus evolves, and more precise decisions for individuals at high risk.

Immunologists call such measurements *correlates of protection.* A correlate is not necessarily a mechanism. It is a marker that travels with protection, the way smoke travels with fire. Sometimes it is the fire itself; sometimes it is merely a dependable sign. The gold standard would be a *mechanistic* correlate: a measurable immune feature that is not just associated with protection but is required for it. In practice, many correlates are statistical: above this level, infections become rare; below it, they become common. The immune system offers many candidates, but turning candidates into reliable signatures is a slow, careful kind of detective work.

The most familiar candidate is antibody. Antibodies are proteins made by B cells that can bind to a virus or bacterium and block it, tag it for destruction, or recruit other immune weapons. In many infections, the antibodies that matter most are *neutralizing antibodies*: those that prevent a pathogen from entering cells. Neutralization is a simple-sounding idea with enormous appeal. It promises a clear threshold: enough neutralizing antibody, and the pathogen is stopped at the door.

There is a reason this idea became so dominant, and it has a history worth remembering. In the late nineteenth and early twentieth centuries, before scientists could see viruses clearly and long before they could sequence genomes, they could still do something powerful: transfer serum (the antibody-containing portion of blood) from an immune animal to a non-immune one, and watch protection appear. Serum therapy saved lives from diphtheria and tetanus and helped establish antibodies as a tangible substance that could carry immunity. Even now, the logic remains seductive: if immunity can be poured from one body into another, surely a measurement in blood should capture much of it.

And often it does. For several vaccines, antibody levels correlate strongly with protection. It is not always the total antibody concentration that matters, but the concentration of antibodies with specific functions. Some must bind the right part of the pathogen; some must neutralize; some must recruit complement or engage immune cells through their Fc region. Laboratories therefore measure antibodies in more than one way: binding assays that count how much antibody recognizes a target, and functional assays that test what those antibodies do.

Yet the immune system is rarely so obliging that one number captures the whole story. For respiratory infections, protection from severe disease may depend heavily on systemic antibodies in blood, while protection from infection itself may depend on antibodies at mucosal surfaces, such as the nose and throat, where pathogens first land. Those mucosal antibodies, often of the IgA class, are harder to measure routinely and may not track neatly with blood measurements. Cellular immunity also matters: T cells can limit disease by killing infected cells and orchestrating responses even when antibodies fail to block infection at the door.

The search for correlates becomes more urgent and more complicated when the pathogen evolves. If a virus changes its surface proteins, neutralizing antibodies generated by vaccination may bind less well. A person might still be protected against hospitalization because T cells recognize more conserved internal viral pieces, but they might get infected more easily. The immune signature of safety may then shift from a simple antibody threshold to a composite picture: some antibody, some T cell function, some speed of recall.

A useful way to think about the problem is to ask what "protected" means at the level of events. Infection is not an on/off switch. It is a race. A pathogen enters. It tries to replicate. The immune system tries to stop it, first with whatever is already in place and then with a rapidly expanding response. Protection from infection means winning very early. Protection from severe disease means winning before tissues are badly damaged. A correlate of protection is, in effect, a laboratory proxy for winning that race.

This is where the single-cell perspective changes the tone of the discussion. Instead of imagining immunity as a single wall of defense, it becomes a collection of capacities distributed across cells with different readiness states. One person may have memory B cells poised to rapidly become antibody-secreting plasmablasts. Another may have tissue-resident memory T cells stationed in the lung. Another may have circulating T cells that proliferate explosively once they see antigen. These differences can lead to similar outcomes, which means that a single correlate may be elusive: there can be multiple immune routes to safety.

So why not simply measure everything? Modern immunology can measure hundreds of features: dozens of cytokines, many antibody specificities, T cell frequencies, transcriptional signatures, receptor repertoires. The obstacle is not measurement alone. It is *interpretability* and *generalisability*. A correlate must work across ages, genetic backgrounds, and health conditions; across viral variants and exposure doses; across laboratories with different assays; across time. A signature that looks impressive in one dataset may fail in another because it captured an accident of study design rather than an underlying law.

A correlate must also be *usable*. Public health decisions cannot wait for a month-long multi-omic analysis on every patient. A good correlate is like a well-calibrated instrument: reliable, affordable, and simple enough to become routine.

The most common statistical approach to correlates starts with a basic question: as a biomarker increases, does the probability of disease decrease? If the relationship is smooth, researchers can fit a curve and identify a region where risk drops sharply. In ideal cases, the curve suggests a threshold. In real life, it often suggests a zone: below it, risk is high; above it, risk is low; in between, outcomes depend on other factors. This is not a failure of the immune system; it is a description of biology interacting with chance.

The hard part is deciding which biomarker deserves the most trust. Neutralizing antibodies are attractive because they can be measured in vitro and because they map to a simple mechanism. But cellular correlates can be equally important, especially when antibodies are incomplete. Measuring T cell immunity is more demanding. T cells are less abundant in blood than antibodies are in serum. They are also more diverse: helper T cells (often $CD4^+$) support B cells and coordinate responses; killer T cells (often $CD8^+$) destroy infected cells. Assays such as ELISpot can measure the frequency of T cells producing a cytokine in response to antigen. Flow cytometry can measure multiple cytokines and activation markers. Newer methods can measure T cell receptor sequences and activation programs at high resolution. Each assay captures a slice of a complex process.

A historical episode that illustrates both promise and complexity comes from HIV. For decades, researchers hunted for a correlate of protection that could guide vaccine design. Neutralizing antibodies against HIV are difficult to generate because the virus mutates rapidly and shields critical sites. Cellular immunity can control viral replication but does not necessarily prevent infection. The HIV experience taught immunology a sobering lesson: for some pathogens, the immune system's usual signatures of safety are hard to achieve, and correlates may be multifactorial or absent. That lesson now shadows every new pathogen: when a correlate appears straightforward, it is wise to test it harshly.

Correlates are also deeply entangled with the immune system's internal safety mechanisms. The immune system must protect without destroying the body. That means it builds in brakes, checkpoints, and rules of engagement. Those rules are not philosophical; they are molecular.

A central rule for T cells is the *two-signal* logic. Signal 1 is antigen recognition: the T cell receptor binds a peptide presented by another cell. Signal 2 is costimulation: additional molecular interactions that say, in effect, "this is worth responding to." Without costimulation, a T cell can become unresponsive or tolerant rather than activated. This logic reduces the risk of responding to harmless or self-derived antigens. It also means that the presence of antigen-specific T cells alone is not enough to guarantee a protective response. A person may carry memory-like cells that recognise a pathogen, yet if the context does not provide the right costimulatory and inflammatory cues, the response may be muted. A correlate that counts antigen-specific T cells but ignores their functional readiness can overpromise safety.

The importance of inhibitory brakes is equally stark. CTLA-4 is one of the best-known inhibitory receptors: it competes with a stimulatory receptor (CD28) for binding to ligands on antigen-presenting cells. When CTLA-4 is absent, the consequences are catastrophic. Experiments in mice lacking CTLA-4 showed severe lymphoproliferation and early death, driven by uncontrolled T cell activation. The lesson is not merely that the immune system can be too weak; it can also be too strong. Any signature of safety must respect this balance. A vaccine that produces enormous activation markers might look impressive in a dataset but could be unsafe if it pushes too hard against the immune system's brakes.

Regulatory T cells, often marked by the transcription factor Foxp3, embody this balancing act. They are not passive; they actively suppress and shape immune responses. Their job is to prevent autoimmunity, limit collateral damage, and maintain tolerance at surfaces exposed to constant stimulation, such as the gut. If one imagines protection as simply more immune activity, regulatory T cells look like a nuisance. In reality, they are part of what makes durable protection possible. An immune system that cannot shut down inflammation will injure the host and, paradoxically, may become less effective over time.

This complicates correlates in a subtle way. Sometimes a good outcome is associated with *less* of a particular inflammatory marker, not more. A person who clears infection efficiently may show a smaller cytokine surge than someone who struggles. In that case, inflammation is a marker of failure, not success. Correlates of protection are therefore not always about high levels; they can be about the right timing and the right restraint.

The idea that context matters was sharpened by Polly Matzinger's "danger theory," which argued that the immune system responds not only to foreignness but to signs of damage and stress. In this view, antigen is not sufficient. The immune system asks, in molecular terms, "Is the tissue in trouble?" Microbial patterns (often called MAMPs) and damage-associated signals from stressed cells (often called DAMPs) act as adjuvant-like cues that license strong responses. This helps explain why harmless proteins can be tolerated while the same proteins mixed with an inflammatory adjuvant become immunogenic. It also helps explain immune evasion: tumors and some viruses can dampen danger signals, quieting the context that would otherwise provoke a robust response.

A correlate of protection, then, might need to reflect not only memory components but the system's ability to be properly alarmed. An immune response that is too easily triggered risks allergy and autoimmunity; one that is too hard to trigger risks infection and cancer. Safety is a decision, not a reflex.

There is also the phenomenon of *exhaustion*, most famously observed in chronic infections and cancer. Exhausted T cells are not simply worn out; they adopt a regulated program characterized by specific transcriptional regulators and inhibitory receptors. This program can be protective in a grim way: it prevents uncontrolled immunopathology when antigen persists. But it also limits the ability to clear pathogens and tumors. For correlates, exhaustion is a warning label. A blood test might show plenty of antigen-specific T cells, but if many of them are in an exhausted state, their protective capacity may be limited. Conversely, a therapy that reinvigorates exhausted T cells, such as checkpoint blockade in cancer, can restore function but also carries risks of autoimmunity. The immune signature that predicts tumor control may not align with the signature that predicts minimal collateral damage.

This brings the discussion back to why correlates are so sought after: they promise prediction without exposure. A correlate is a kind of *immune weather forecast*. It does not say what will happen with certainty; it says what is likely under certain conditions. Like weather, immunity depends on many interacting variables, and a single thermometer reading is rarely enough. Still, some measurements are so informative that they become standard.

A concrete example is hepatitis B vaccination. For years, antibody levels to the hepatitis B surface antigen have been used as a practical marker of protection. The presence of a sufficiently high antibody titer after vaccination generally correlates with protection, and this has shaped clinical guidelines, especially for healthcare workers and immunocompromised patients. The success of this correlate has made it a model: find the antibody threshold, and you can decide who needs boosting. Yet even here, immunology keeps a footnote ready. Antibody levels can wane while immune memory persists via memory B cells; a person with low antibodies may still mount a rapid response upon exposure. The correlate works well enough for policy, but it is not a perfect metaphysical definition of safety.

The best correlates are built through a chain of evidence. First, a biomarker is measured in many individuals. Second, outcomes are recorded: infection, disease severity, transmission. Third, statistical models link biomarker levels to risk. Fourth, the relationship is tested in new cohorts and across laboratories. Finally, mechanistic experiments support the causal interpretation: antibodies from high-titer individuals neutralize the pathogen; T cells from protected individuals show functional capacity; interventions that change the biomarker change protection. Each step reduces the chance that the correlate is merely coincidental.

There is a human dimension to this work that is easy to miss in graphs. A correlate of protection can speed vaccine access in a crisis, but it can also exclude people if misapplied. If the correlate is based on data from young healthy adults, it may not predict well for older adults, whose immune systems often respond differently. If the correlate is derived from one population, it may not generalize to others with different prior exposures, genetics, or health burdens. Immune memory is shaped by a lifetime of infections and vaccinations; two people can reach similar antibody levels through very different immunological histories. A good signature of safety must be robust to those differences, or at least honest about its limits.

The most mature view treats correlates as part of a negotiation between biology and decision-making. Biology offers patterns with error bars. Decision-making demands thresholds: who gets a booster, who can return to work, which vaccine candidate advances. The correlate is where these worlds meet. It is a compromise that aims to be accurate enough to guide action while remaining humble enough to be revised.

The most reassuring feature of a strong correlate is not that it never fails, but that when it fails, it fails in predictable ways. If protection is high above a threshold but not guaranteed, that is honest. If certain groups systematically deviate, the correlate can be adjusted or stratified. The immune system is too complex for a single signature to be eternal, but it is not too complex for useful signatures to exist.

In the end, the search for correlates of protection is less like finding a magical seal and more like learning the immune system's handwriting. The signature of safety is not a single flourish. It is the particular combination of marks that, across many lives and many exposures, reliably means: the body has practiced this before, it knows what to do, and it can do it fast enough to matter.

9.3 Circuit Boards of Biology

An immunologist can hold a vial of cells that, on paper, are all the same: "CD8 T cells," "memory phenotype," perhaps even members of the same clone. Add a pinch of antigen, a dash of cytokine, and the vial changes character. Some cells divide furiously; some release toxic granules; some become quiet and long-lived; some edge toward exhaustion; some die. If the immune system were a simple machine, the same input would yield the same output. Instead it behaves like a complicated device with hidden wiring: a network of sensors, relays, amplifiers, and safety interlocks. The phrase "black box" is not an insult; it is a confession that the outcome can be predicted only loosely from the ingredients.

Computational immunology is the attempt to open the casing without destroying the device. It does not begin with the fantasy that a cell is a tiny person deliberating. It begins with a more modest idea: a cell is an information-processing system. Signals arrive at the surface, are translated into chemical changes inside, and ultimately alter gene expression and behaviour. Information processing is something engineers have studied for a long time, and the language they use—inputs, outputs, circuits, logic gates—turns out to be surprisingly good at describing what immune cells do.

A logic gate is a simple rule that produces an output only when particular combinations of inputs are present. The classic textbook gate is AND: output happens only if input A *and* input B are both present. Immunology has gates everywhere. A T cell typically requires antigen recognition *and* costimulation to fully activate. A B cell's decision to become a high-affinity antibody producer is shaped by antigen binding *and* help from T cells. A memory cell's choice to proliferate depends on antigen *and* inflammatory context, tempered by inhibitory receptors that function like NOT gates. Real biology is analogue rather than digital, with graded signals and noisy components, but the gate metaphor captures something crucial: immune responses are conditional. The system is built to prevent accidental activation, and it therefore demands combinations.

The shift from metaphor to model has been accelerated by two kinds of data. The first is the detailed single-cell measurements that reveal states and transitional trajectories rather than averaged populations. The second is perturbation: the ability to push specific buttons—knock out a gene, activate a receptor, block a pathway—and watch the system respond. Data without perturbation is like watching a city from above: you can see traffic patterns but not the rules that produced them. Perturbation without single-cell resolution is like pushing a button and only measuring the city's average speed. Put them together and the wiring begins to show.

It helps to remember how different this is from older traditions in immunology. For much of the twentieth century, the field was organized around parts lists: which cells exist, which molecules they express, which cytokines they produce. That was necessary groundwork. But parts lists do not explain behaviour. Knowing every component in a radio does not tell you how stations are tuned. The turning point in many areas of biology came when researchers stopped asking only “what is there?” and began asking “what connects to what?” and “what happens if I cut this wire?”

A concrete and familiar example of immune circuitry is the response of a dendritic cell to bacterial lipopolysaccharide (LPS), a component of the outer membrane of many bacteria. LPS is an alarm signal recognized by Toll-like receptor 4. When dendritic cells encounter LPS, they change from quiet sentinels into activated teachers: they upregulate costimulatory molecules, secrete cytokines, and prepare to prime T cells. This transformation is not a single switch; it is a cascade. Some genes turn on within minutes; others hours later. Some programmes are pro-inflammatory; others are regulatory brakes that prevent runaway damage. From far away, one might call this “activation.” Up close, it looks like a sequence of logic gates and feedback loops: IF LPS, THEN NF-κB activation; IF NF-κB AND certain chromatin states, THEN cytokine genes; IF cytokines are produced, THEN autocrine signalling that reinforces or reshapes the programme; IF inflammation persists, THEN induction of negative regulators that dampen the response.

The first step toward turning such narratives into computation is to admit that single-cell data are snapshots. A cell is captured at one moment in time, frozen into a data point. Yet immune decisions are dynamic. The question becomes: can one infer motion from still images?

RNA velocity, introduced by Gioele La Manno and colleagues (2018), offered a striking solution. Newly transcribed RNA exists in an “unspliced” form before being processed into mature “spliced” mRNA. Many single-cell protocols capture both forms. The ratio of unspliced to spliced reads for a gene carries information about whether that gene is being turned up or down. If unspliced RNA is abundant relative to spliced, transcription is rising; if unspliced is scarce, transcription may be slowing. By aggregating these gene-level hints, RNA velocity estimates a direction of change for each cell in gene-expression space. It does not predict the future with certainty, but it adds an arrow to the dot.

When applied to immune systems, this is immediately intuitive. During an infection, a naive T cell does not leap instantaneously into a mature effector cell; it passes through activation states, proliferative bursts, and differentiation programmes. A single-cell snapshot might show a cloud of cells that seem to grade smoothly from naive to effector. Velocity adds a sense of which way the crowd is flowing and where branches might occur: a subset heading toward long-lived memory, another toward short-lived effector fate, another toward exhaustion under persistent stimulation.

Another set of tools treats the data not as points on a map but as a network. PAGA, developed by F. Alexander Wolf and colleagues (2019), builds a graph connecting clusters of cells and estimates the confidence of transitions between them. This is helpful because the immune system is both discrete and continuous. Some categories are real: a B cell is not a macrophage. But within T cells, many "types" are better understood as regions of a continuum. Graph abstractions respect both: they allow islands connected by bridges, not just islands separated by water. In practical terms, this helps interpret a single-cell dataset from an infection or vaccine: are there truly separate lineages, or are we imposing labels onto a smooth developmental landscape?

Trajectory inference, however, still leaves a deeper puzzle. Even if one can trace the path from naive to memory, what determines the turnings? At a branching point, why does one cell become a tissue-resident memory cell while another becomes a circulating memory cell? Why does one adopt an exhausted programme while another retains effector capacity? Trajectories describe *what* happens. Circuits aim to explain *why*.

Here the language of gene regulation becomes central. Inside each cell, transcription factors act as molecular decision-makers. They bind DNA, recruit machinery, and turn sets of genes on or off. A transcription factor rarely acts alone. It operates in a network: transcription factors regulate each other, collaborate, and compete. The result can resemble modules: coherent programmes that switch on together, producing recognizable states.

SCENIC, developed by Sara Aibar and colleagues (2017), is one approach to inferring such networks from single-cell RNA sequencing. Rather than clustering cells based simply on expression of individual genes, it attempts to identify *regulons*: transcription factors together with their predicted target genes. A regulon is a hypothesis about wiring. If the regulon is active, the cell is likely operating under that transcriptional programme. This shift is conceptually important. It moves the analysis from descriptions ("these cells express gene X") to mechanisms ("these cells appear to be controlled by transcription factor Y"). Regulon activity can sometimes look more discrete than raw gene expression, suggesting that cells may switch between regulatory regimes in a relatively abrupt way even when their transcriptomes look continuous.

At this point, the analogy to circuit boards becomes more than decorative. Many circuits, both electronic and biological, are built from motifs: feed-forward loops that filter noise, positive feedback loops that create stable states, negative feedback loops that prevent runaway. The immune system uses all of these. Consider the need for decisiveness. A T cell that is half-activated is dangerous: it may produce some inflammatory molecules without fully committing to a controlled programme. Positive feedback, such as autocrine IL-2 signalling reinforcing proliferation, can push cells toward a committed state. Consider the need for restraint. The same proliferation must stop. Negative feedback, such as induction of inhibitory receptors and regulatory cytokines, provides the braking system.

But inferred networks remain correlational unless the wiring is tested. Correlation is not causation; it is not even direction. Does transcription factor A activate programme B, or does programme B induce transcription factor A, or are both downstream of an unseen signal? To move from a guessed circuit diagram to a reliable one, biologists need interventions.

Perturb-seq, introduced by Jonathan Dixit, Adamson and colleagues (2016), is a particularly elegant marriage of CRISPR gene perturbation and single-cell RNA sequencing. The idea is simple to state and transformative in effect. Use CRISPR to knock out or perturb genes in a pooled population of cells, but include a barcode so that, when you sequence each cell, you can tell which perturbation it received. Then read out the entire transcriptome. Each cell becomes a data point with a known intervention. Instead of merely observing that gene X and gene Y are co-expressed, one can ask: if I remove gene X, how does the whole programme change? Which pathways collapse, which compensate, which alternative fates appear?

In immune contexts, perturb-and-readout approaches can map the regulatory circuits controlling activation, cytokine production, antigen presentation, and differentiation. Returning to the LPS example, one can perturb dozens or hundreds of candidate regulators and see which ones are required for the early inflammatory burst, which ones for later interferon responses, and which ones for negative regulation. It is like systematically disabling components on a circuit board and watching which lights go out.

One might wonder why such systematic mapping has not always been done. The main answer is scale. Immune circuits are large. They involve many genes, and the combinations grow explosively. If you perturb one gene at a time, the space is big but manageable. If you perturb pairs to detect interactions, the number of experiments becomes enormous. Newer designs, sometimes called compressed perturbation screens, attempt to handle this combinatorial explosion by using clever pooling schemes and computational reconstruction, exploiting the fact that many circuits are sparse: not everything interacts with everything. The details can be technical, but the direction is clear. The field is moving from artisanal circuit sketches to industrial mapping.

Computational modelling does not stop at mapping. The deeper ambition is simulation: if the wiring is known, can one predict what a cell will do under new combinations of signals? This is where logic gates return. A model can represent pathways as conditional rules. For example: IF antigen is present AND IL-2 signalling is above a threshold AND inhibitory signalling is low, THEN proliferate. IF antigen is persistent AND certain stress programmes are active, THEN shift toward exhaustion. Real models often use differential equations, probabilistic frameworks, or machine learning, but the conceptual heart is similar: encode dependencies so that, given inputs, the model outputs likely fates.

Simulation has a special appeal in immunology because the system is hard to explore exhaustively in the laboratory. Signals arrive in combinations, sequences, and doses that are difficult to recreate. Human tissue sampling is limited. Animal models capture some features and miss others. A good in silico model offers a sandbox: millions of hypothetical scenarios can be tested cheaply, and the most informative can be selected for real experiments. It is not a replacement for biology; it is a way of spending biological effort wisely.

The dream becomes particularly vivid when one considers the engineering of immune cells in medicine. Cancer immunotherapy has already begun to treat immune circuits as things that can be rewired. Chimeric antigen receptor (CAR) T cells are a famous example: T cells are genetically modified to express a synthetic receptor that recognises a tumour antigen and triggers activation. The therapy is powerful and sometimes dangerous, in part because the activation circuit can be too strong or too indiscriminate. Newer generations of engineered cells incorporate more explicit logic: receptors that require two antigens (an AND gate) to reduce off-tumour toxicity, or inhibitory receptors that block activation in certain tissues (a NOT gate). Even without naming specific platforms, the conceptual point stands: once immune decisions can be expressed in logic, they can be redesigned with safety constraints in mind.

This engineering perspective also clarifies what "decoding" really means. The immune system is not a single circuit board. It is many boards communicating: dendritic cells sensing and presenting; T cells integrating antigen and context; B cells refining antibodies; tissue cells sending distress signals; regulatory cells enforcing restraint. Each board has its own wiring, but the boards are coupled. A computational model that focuses on one cell type must decide what it treats as input from the environment and what it models explicitly. This is not a flaw; it is the same challenge faced by any engineer modelling a subsystem in a complex machine.

Cross-dataset integration has therefore become essential. A circuit diagram that only works for one dataset, generated in one laboratory under one protocol, is not a diagram of biology; it is a diagram of an experiment. Methods for integrating single-cell datasets across batches, platforms, and conditions, such as anchoring approaches in tools like Seurat (Stuart and colleagues, 2019), aim to build reusable reference maps: a stable coordinate system where a new dataset can be placed, and where cell states have consistent meaning. In immune memory, this matters because memory is shaped by life history. To understand what is universal and what is contingent, one must compare across people, pathogens, and time.

The most honest way to describe the current state of computational immunology is to say that it has made immune decisions *legible* without making them fully predictable. The models are not crystal balls. They are structured guesses constrained by data and tested by perturbation. Their success is often local rather than global: excellent predictions in a defined context, weaker predictions when the context shifts. That is not a disappointment; it reflects a fundamental feature of immunity. The system evolved to be adaptable. Adaptability is, by definition, hard to compress into a fixed formula.

Still, there is a kind of clarity that arrives when circuitry becomes the language. A memory cell is no longer a mysterious unit of "stored experience." It is a physical system with thresholds, feedback, and regulatory programmes. It can be tipped into action or held back. It can be rewired by persistent antigen, chronic inflammation, or therapeutic intervention. When a vaccine works, it does not simply "teach" the immune system; it builds and tunes circuits so that later, when a pathogen appears, the right gates open quickly enough and the brakes engage before damage becomes self-inflicted.

That image has consequences beyond the laboratory. It suggests that immune safety is not merely a matter of having more immunity, but of having immunity whose wiring has been tuned to the right conditions: strong enough to respond, selective enough to avoid friendly fire, flexible enough to adapt to variation, and restrained enough to preserve the tissues it exists to defend.

Chapter 10

Engineering the Future

We are crossing the historical threshold from merely describing immune behavior to actively rewriting the code of life. By deciphering the molecular handshakes that determine whether a cell chooses tolerance or aggression, we gain the power to untrain autoimmune errors and unleash targeted fury against hidden tumors. This chapter examines how the signaling logic discussed previously is now being hacked, tweaked, and reprogrammed to create living therapies.

10.1 The Art of Releasing the Brakes

Cancer is often described as a problem of disguise: malignant cells "hide" from the immune system by blending in. That picture is comforting, because it suggests a simple fix—make the cancer visible, and immunity will do the rest. The unsettling truth is that many tumors are not especially invisible. They can be recognized. T cells may even be sitting right at the tumor's edge, already holding the molecular memory of what they have seen. And yet nothing happens.

To understand why, it helps to stop thinking of the immune system as a set of weapons and start thinking of it as a system of permissions. A T cell is not a bullet you fire; it is a decision-making cell that must be convinced, step by step, that killing is justified. The same features that make immunity precise—its reluctance to attack without sufficient evidence—also create a vulnerability. A tumor does not need to erase every sign of its abnormality. It can survive by leaning on the immune system's own caution.

That caution exists for good reasons. A T cell's job description is terrifying: find a cell in the body and destroy it. If that power were unleashed casually, autoimmunity would be the default state of life. So evolution built in braking systems: molecular "stop signals" that dial down activation once danger seems to have passed, and that prevent T cells from attacking in ambiguous situations. These brakes are not design flaws; they are essential safety features. They are also exploitable.

Checkpoint inhibitor drugs, among the most consequential advances in modern oncology, work by interfering with those brakes. They do not directly poison tumors. They do not (usually) make tumors more visible. Instead, they remove a layer of restraint that has been suppressing an existing immune response. In many patients who benefit, the immune system was already in the right neighborhood, asking the right questions, and being told—persistently, chemically—to stand down.

The phrase "immune checkpoint" can sound like airport security, but the metaphor is apt. At a checkpoint, you do not pass just because you exist; you pass because your documents satisfy a set of conditions. In the immune system, one of the most important documents is an antigen: a molecular fragment that signals what a cell is making. T cells inspect these fragments when they are displayed on the surface of other cells, held in place by proteins called MHC (major histocompatibility complex). Recognition is necessary, but it is not enough. A T cell also listens for context. Is there evidence of infection or tissue damage? Are supportive signals present from other immune cells? Or are there explicit inhibitory messages insisting that restraint is safer?

Two checkpoint pathways became famous because tumors use them so effectively: CTLA-4 and PD-1. Their names are technical, but their logic is simple.

CTLA-4 acts early, mostly during the first moments when a T cell is being activated by a professional antigen-presenting cell such as a dendritic cell. The activation of a T cell requires not only recognition through its T-cell receptor (TCR) but also a "go" signal called costimulation. One of the central costimulatory receptors is CD28, which binds to CD80 and CD86 on the antigen-presenting cell. CTLA-4 is a close cousin of CD28 that competes for those same ligands and, in effect, steals the microphone. In a striking mechanistic twist uncovered years after CTLA-4 entered the spotlight, CTLA-4 can physically remove CD80 and CD86 from the opposing cell by pulling them across the contact zone and internalizing them, reducing the ability of that antigen-presenting cell to provide costimulation. The T cell is not merely hearing "no"; it is dismantling the very apparatus that would have helped it hear "yes."

PD-1 (Programmed cell death protein 1) operates more in the periphery, where T cells encounter their targets in tissues. It is upregulated after activation and becomes prominent when stimulation is chronic—the kind of prolonged engagement that happens in many cancers. PD-1 binds to PD-L1 or PD-L2 on other cells. When PD-1 is engaged, the T cell's internal signaling is rewired toward restraint. A useful way to picture PD-1 is not as a gag on the T cell's mouth, but as a hand on its steering wheel: it biases downstream signals away from sustained attack, particularly by dampening costimulatory pathways that help T cells remain effective.

Tumors exploit this in a perverse imitation of normal physiology. In healthy tissues, PD-L1 expression can protect against collateral damage during inflammation. In pregnancy, similar tolerance mechanisms help prevent immune attack on the fetus. In infections, they help prevent an immune response from overshooting and destroying the host. Many tumors, however, express PD-L1 as a shield. Some do it constitutively because of oncogenic signaling. Others do it adaptively: when T cells arrive and release interferon-γ (a cytokine associated with immune attack), the tumor responds by increasing PD-L1, turning the immune assault into the trigger for its own protection. The immune system is, in a sense, talking itself into silence.

It is hard to overstate the conceptual leap that checkpoint therapy represented. For decades, much of cancer immunotherapy focused on adding more: more immune cells, more cytokines, more stimulation. Those approaches sometimes worked, but they were often like trying to win an argument by shouting. Checkpoint blockade tried something psychologically different: it assumed the immune system might already be shouting, and asked whether something was forcing it to whisper.

A historical anecdote captures the shift. In the mid-1990s, work in James P. Allison's circle helped crystallize CTLA-4 as an inhibitory receptor rather than an activating one, and preclinical experiments by Leach, Krummel, and Allison (1996) showed that blocking CTLA-4 with an antibody could enable mice to reject tumors and then resist re-challenge—a functional sign of immune memory. The detail that matters is not the mouse model but the logic: an antibody that does not bind the tumor at all, but binds a T cell regulator, could produce durable protection. Tumor rejection was not merely a transient flare of inflammation; it could leave behind a memory trace, like an immunological "wanted poster" that made subsequent encounters easier to handle.

The later clinical triumphs—with CTLA-4 blockade and, especially, PD-1/PD-L1 blockade across multiple cancers—did not come from teaching the immune system a brand-new lesson in every case. They often came from permitting an old lesson to be acted upon.

That "double negative" logic is worth slowing down for. A T cell has activating pathways (accelerators) and inhibitory pathways (brakes). Tumors frequently do not disable accelerators directly; they press the brakes. A checkpoint inhibitor is an inhibitor of inhibition. It blocks the brake. The result is not an artificially imposed attack, like turning on a machine. It is more like taking your foot off the brake pedal and allowing an engine that is already running to move the car.

The decisive action takes place at the immune synapse: the intimate contact zone where a T cell and its counterpart cell (a tumor cell, or an antigen-presenting cell) exchange information. This synapse is not a single handshake but a crowded marketplace of molecular touches: recognition, adhesion, costimulation, inhibition. PD-1 and PD-L1 form one of the most clinically important inhibitory handshakes.

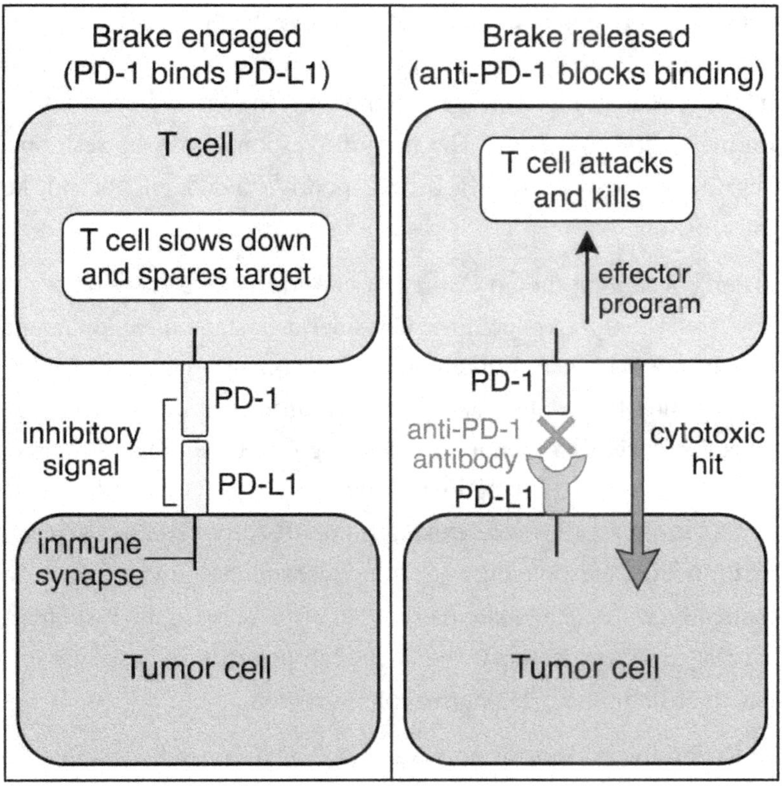

This diagram hides a crucial subtlety: blocking PD-1 does not guarantee that a T cell will attack appropriately. It merely removes one strong reason not to. Whether the cell becomes an effective killer depends on the rest of the conversation at the synapse.

To see why, imagine a T cell as a highly trained professional who will not act on a single clue. Recognition of antigen is like matching a fingerprint, but costimulation is like confirming jurisdiction and authorization. Without that confirmation, the immune system tends to interpret recognition as ambiguous or even normal. A tumor microenvironment can be rich in ambiguity: low nutrients, suppressive cytokines, metabolic byproducts, and immune cells whose job is to keep inflammation in check. If you remove the brake in a car whose engine is broken, you will not go far. That is one reason checkpoint inhibitors can produce dramatic remissions in some cancers and frustratingly little in others. The immune response may be restrained, yes; it may also be poorly fueled, poorly guided, or missing key information.

When checkpoint therapy works, it can look like a story with a delayed but satisfying punchline. Consider metastatic melanoma, once a diagnosis with very limited durable treatment options. Melanoma is often highly mutated, and those mutations can create novel antigens that T cells can recognize as foreign. Yet the tumor's microenvironment can be saturated with inhibitory signals. Blocking PD-1 or CTLA-4 can allow pre-existing tumor-reactive T cells to expand and function. In some patients, tumors shrink not gradually but dramatically, as if a stalemate has ended. And because the responding T cells can form memory, the effect can persist long after the drug has done its immediate pharmacological work.

But releasing brakes has consequences. The same checkpoints that protect tumors can also protect healthy tissues. If the drug reduces restraint broadly, the immune system may redirect its newly liberated energy toward normal organs. Clinicians call these immune-related adverse events. The list can read like a tour of the body: colitis (inflammation of the colon), dermatitis, hepatitis, pneumonitis (lung inflammation), thyroid dysfunction, hypophysitis (inflammation of the pituitary), arthritis-like symptoms, and more. The common thread is not the organ but the logic: a safety mechanism has been loosened.

This is not simply "too much immunity." It is misallocated permission. The immune system is being encouraged to treat some signals as more dangerous than it otherwise would. In many cases, the adverse events are manageable with careful monitoring and immunosuppression (often corticosteroids), but the need for these countermeasures reveals what checkpoint therapy truly is: a controlled experiment in tolerance. You are adjusting the immune system's rules of engagement, not aiming a toxin at a tumor.

One reason PD-1 blockade was such a conceptual breakthrough is that it made immunological exhaustion clinically visible. In chronic infections and cancers, T cells can enter a state called exhaustion, where they remain present but have reduced function and altered gene expression. Exhaustion is not laziness; it is a protective adaptation to constant stimulation. PD-1 is one marker and mediator of this state, though it is not the whole story. If exhaustion were merely a reversible pause, removing PD-1 signaling might always restore vigor. In reality, some exhausted states become partly entrenched, and some tumors evolve to evade immune pressure in other ways: losing antigen presentation machinery, altering interferon pathways, or building a microenvironment that physically excludes T cells. Checkpoint inhibitors are powerful, but they do not rewrite every constraint.

Still, the idea that therapy can succeed by interrupting inhibition rather than adding aggression has spread far beyond PD-1 and CTLA-4. It changed what oncologists look for and how patients talk about their own disease. A patient responding to a checkpoint inhibitor is not simply "taking a drug." They are participating in an altered negotiation between their immune memory and their tumor's attempts at self-protection. The tumor becomes less like an invading army and more like a corrupt diplomat, using the host's rules to secure immunity from prosecution. The drug is not a new weapon; it is a revocation of diplomatic immunity.

There is a final, paradoxical comfort in this story. Checkpoint therapy is sometimes framed as an attempt to make the immune system do something unnatural: attack the body. In fact, it often reveals that the immune system has been trying to do something natural all along: eliminate dangerous cells while avoiding unnecessary harm.

The tragedy of cancer is that those two goals can come into conflict, and tumors learn to stand precisely where the immune system is most reluctant to act. The art of releasing the brakes is learning when restraint has become the greater danger, and when a carefully placed interruption in a single molecular handshake can allow immune memory to finish an argument it started long ago.

10.2 Restoring the Truce

If cancer therapy sometimes succeeds by taking the immune system's foot off the brake, autoimmune disease poses the opposite dilemma: the immune system is already pressing the accelerator against the body's own tissues. In rheumatoid arthritis, fingers stiffen because immune cells treat joints like persistent irritants. In multiple sclerosis, the insulating myelin around nerve fibers becomes a target, and the consequences can be weakness, numbness, blurred vision, and fatigue that is not merely tiredness but an oppressive gravity. In type 1 diabetes, immune cells destroy the insulin-producing β cells of the pancreas, turning a finely tuned metabolic reflex into a daily calculation of risk.

The medical triumph of the last half-century has been learning to dampen these attacks. Steroids, anti-inflammatory drugs, and targeted biologics can dramatically improve lives. Yet almost all of them share an uncomfortable trait: they blunt immunity in broad strokes. They quiet the immune system the way you might quiet an orchestra by turning down the electricity to the whole building. The music gets softer, but so does the fire alarm. Patients trade symptoms of autoimmunity for greater susceptibility to infection, slower wound healing, and sometimes a higher risk of certain cancers.

The dream behind "inverse vaccines" is more daring and more humane. A normal vaccine teaches the immune system to remember a threat. An inverse vaccine aims to teach it *not* to respond—to create a selective, antigen-specific calm. Instead of "here is a pathogen; prepare to fight," the message becomes "here is a familiar molecule; stand down." The hope is to restore a truce: not an armistice achieved by disarming the entire immune system, but a precise treaty that applies to one set of molecular identities.

Calling this "amnesia" is partly metaphorical. The immune system does not erase memories the way a computer deletes files. But it can be trained into forms of tolerance that, in practice, resemble forgetting: the relevant T cells become inactive, or are steered into a more regulated state, or are outcompeted by regulatory cells that actively suppress harmful responses. The immune system learns, again, that certain targets are not enemies.

To make sense of inverse vaccines, it helps to understand what tolerance really is. Tolerance is not simply the absence of immunity; it is a *maintained state*. The immune system is filled with cells capable of recognizing the body's own molecules. Some are eliminated during development, but not all. Those that remain are managed by a web of rules: the requirement for the right context, the presence of inhibitory signals, the actions of specialized "peacekeeping" cells. Among the most important peacekeepers are regulatory T cells, often abbreviated as Tregs. They are not passive. They patrol, interpret, and suppress. They use chemical messages and cell-to-cell contact to dampen overzealous responses, and they help prevent an everyday catastrophe in which immunity would turn ordinary life into chronic inflammation.

Autoimmune disease emerges when these systems fail. Sometimes the trigger is genetic: variants in immune regulation shift the balance toward attack. Sometimes the trigger is environmental: an infection or injury provides inflammatory context in the wrong place at the wrong time. Sometimes the target is exposed in an unusual way: tissue damage releases proteins that are normally hidden from immune inspection, inviting suspicion. Often it is all of these at once. What matters for therapy is that the immune system's decision-making has been nudged into a new equilibrium, one where the wrong interpretation repeats itself.

A standard vaccine is built from two conceptual ingredients: an antigen (the "what") and a danger signal (the "why now"). The antigen is the fragment that immune cells learn to recognize. The danger signal is the context that tells the immune system the antigen is associated with threat. Infections naturally provide danger signals through molecules that are common in microbes and rare in humans, or through signs of cell damage. Without those danger signals, antigens tend to be tolerated rather than attacked, especially when they are encountered repeatedly and calmly.

An inverse vaccine flips the emphasis. The antigen is still central—it must be, if specificity is the goal—but the crucial design problem becomes: how do you present that antigen in a way that reliably communicates safety?

There is an old, slightly odd historical thread here, one that predates modern immunology. In the early twentieth century, before the immune system's molecular language was mapped out, clinicians noticed that gradual exposure to allergens could reduce allergic reactions. Allergy desensitization was not a cure in the sense of removing the allergen from the world; it was a retraining of the immune response. Over time, many patients reacted less violently to the same substance. The approach was empirically developed, sometimes perilously, and it worked better for some allergies than others. Yet conceptually it offered a clue: immune responses are not fixed reflexes. They can be reprogrammed by the way an antigen is delivered, the dose, the timing, and the surrounding immune environment.

Autoimmunity is not allergy, but the lesson carries. If the immune system can be taught to become less reactive to pollen or venom, might it be taught to become less reactive to insulin, myelin, or joint proteins?

Modern inverse vaccines pursue this idea with a more precise toolset. The antigen is chosen deliberately, often as short peptides—small fragments of a protein—that are known, or suspected, to be targeted by autoreactive T cells. The delivery is engineered to bias the immune system toward tolerance. The readouts are measured in immune cell subsets, cytokines, gene expression, and—most importantly—clinical outcomes.

Consider type 1 diabetes, where the immune attack is directed at the pancreatic β cells. The tragedy is made sharper by time: by the time diabetes is diagnosed, much of the β-cell mass has already been damaged. Many people can still produce a small amount of insulin at diagnosis, measurable as C-peptide (a byproduct of insulin production). Preserving that residual function matters. Even modest remaining endogenous insulin can make blood sugar easier to control and reduce long-term complications.

In recent-onset disease, researchers have explored antigen-specific immunotherapies that deliver multiple β-cell-related peptides to the immune system in ways designed to calm, not inflame. In at least one genotype-selected randomized placebo-controlled study in people with recent-onset type 1 diabetes, a multi-peptide approach showed immunologic shifts consistent with increased regulatory programs, and a subset of participants showed preservation of C-peptide at around 24 weeks. That is not a definitive victory—autoimmune disease rarely offers clean wins—but it illustrates the style of progress: small, measurable nudges in immune decision-making, paired with cautious clinical signals, and an ongoing struggle to identify who benefits and why.

Multiple sclerosis (MS) provides another revealing test case. In MS, the immune response targets components of myelin, though the precise antigenic targets may differ among individuals and across time. This creates a practical challenge that is less dramatic than engineering a drug but just as stubborn: you cannot tolerize the immune system to a target if you do not know which target matters for a given person at a given stage of disease. Trials of antigen-specific tolerance strategies in MS have varied widely in antigens, routes, and endpoints. Systematic reviews tend to converge on a sober assessment: these approaches often appear generally safe, but evidence for clear clinical efficacy remains ambiguous because the trials are heterogeneous and the disease itself is variable. That ambiguity is not a verdict against the concept so much as a reminder of how hard it is to retrain a complex system when you are not sure which lesson it is currently rehearsing.

The same logic applies to rheumatoid arthritis (RA), where multiple antigens and multiple inflammatory loops can sustain disease. Here, too, antigen-specific strategies have been tested, including approaches using tolerogenic dendritic cells.

Dendritic cells deserve a moment in the spotlight because they are the immune system's professional educators. A dendritic cell picks up proteins, chops them into fragments, and presents those fragments to T cells, alongside contextual signals that encourage either activation or restraint. When dendritic cells are exposed to inflammatory cues, they mature into potent activators. When they are kept in a calm state—or deliberately engineered to be calm—they can promote tolerance.

A tolerogenic dendritic cell, often shortened to TolDC, is essentially a dendritic cell trained (or built) to teach peace. In laboratory protocols, dendritic cells can be generated from a patient's blood cells and conditioned with drugs or cytokines that bias them toward a regulatory phenotype. They can then be "loaded" with specific antigens and returned to the patient. The aspiration is exquisite specificity: show the immune system a particular self-antigen, but present it with the tone and posture of safety. Early clinical programs have treated tolerance as something like a programmable output, influenced by details that might sound mundane but are biologically decisive: the route of administration, whether the cells reach particular lymph nodes, and whether they encounter inflammatory signals along the way.

What, mechanistically, might an inverse vaccine do inside the body?

One possibility is *anergy*: a state in which a T cell remains alive but becomes unresponsive, like a guard who has been ordered to ignore a particular face no matter what alarm is sounded. Anergy can be induced when a T cell recognizes antigen without the proper costimulatory "go" signals. This is one reason the immune system is cautious about activation: the presence or absence of costimulation is used as a proxy for danger.

A second possibility is *deletion*: the physical elimination of the problematic cells. In some contexts, repeated stimulation without appropriate survival cues can lead to activation-induced cell death. Deletion is the closest the immune system comes to erasing a memory, but it is not always achievable and may not be stable if new autoreactive cells continue to arise.

A third possibility is *regulation*: shifting the balance toward Tregs or other regulatory mechanisms that actively suppress harmful responses. Rather than removing the aggressive cells, the immune system recruits more peacekeepers and gives them authority. In practice, this can be a more robust form of tolerance because it can adapt to fluctuations and suppress bystander activation in inflamed tissue.

These outcomes are not mutually exclusive. They also depend on the antigen and on the immune state of the person receiving the therapy. A patient with newly diagnosed autoimmunity may have a different immunological landscape from someone with decades of disease, where inflammation has remodeled tissues and memory cells have become entrenched.

The design of inverse vaccines therefore resembles diplomacy more than engineering a simple molecule. Dose matters: too little antigen may be ignored; too much may inflame. Timing matters: delivering a tolerogenic signal in the middle of an infection may fail because the entire immune system is in a danger-reading mode. Route matters: antigen delivered to the skin or muscle may be interpreted differently from antigen delivered intravenously, orally, or to mucosal surfaces, because those compartments have evolved different default attitudes toward the world. The gut, constantly exposed to food proteins and microbes, has a special talent for tolerance, but exploiting it reliably is difficult.

This complexity can make inverse vaccines sound speculative, but the field has accumulated a quiet confidence about one central claim: antigen-specific tolerance is not fantasy. It is part of normal physiology. The immune system tolerizes to food antigens, to commensal microbes, and to many of the body's own proteins. The problem is not whether tolerance exists, but whether it can be induced deliberately and predictably in the teeth of an ongoing autoimmune response.

That "in the teeth of" is not rhetorical. Established autoimmunity is not merely a mistaken belief; it is a self-reinforcing ecosystem. Damaged tissue releases signals that attract immune cells. Immune cells release cytokines that recruit more immune cells and alter local metabolism. Antigen presentation can broaden, as new self-proteins are revealed by damage, a phenomenon sometimes called epitope spreading. The result is that even if you could tolerize the immune system to one antigen, other antigens might continue the attack. A single treaty may not end a multi-front war.

This is where the most modern inverse vaccine designs begin to look less like a single injection and more like an attempt to control context. Some strategies package antigens into nanoparticles designed to be taken up by specific antigen-presenting cells associated with tolerance. Some aim to deliver antigens alongside molecular cues that mimic the body's own "calm" signals, encouraging dendritic cells to adopt a tolerogenic posture. Others combine antigen-specific approaches with temporary, modest immunomodulation—not a permanent suppression, but a brief window in which inflammatory noise is reduced enough for tolerogenic education to take hold.

There is also a pragmatic ethical motive driving the work. Autoimmune diseases often begin in youth and persist for decades. A therapy that replaces lifelong broad immunosuppression with a period of targeted immune retraining would change not only symptoms but identity: fewer infections, fewer hospital visits, fewer compromises. It would change pregnancy planning, travel, employment, and the everyday feeling of living in a body that cannot always be trusted.

At the same time, precision tolerance carries a different kind of risk. A drug that suppresses immunity broadly is blunt, but predictable in its general direction. A therapy that aims for antigen-specific tolerance must be right about the antigen. If you tolerize the wrong target, you may accomplish nothing. If you tolerize a target that also matters for fighting infections or surveilling cancers, you could create a dangerous blind spot. Biology rarely offers perfectly isolated levers.

The field's cautious tone, then, is not lack of imagination. It is respect for the immune system's depth. Immune memory is not a filing cabinet of "self" and "non-self." It is a dynamic set of preferences shaped by context, tissue, and time. Inverse vaccines try to speak that language fluently: to present the same antigen in a different accent, under different lighting, with a different social cue, until the immune system's conclusion changes from "attack" to "ignore" or, better, "protect."

A truce is not a surrender. It is an agreement to stop treating a familiar neighbor as an intruder. When it works, the victory is not dramatic tumor shrinkage on a scan, but something more intimate: stable blood sugar with less volatility, a hand that opens without pain, a morning that begins without fear of the body's own reflexes. The immune system, so often imagined as a soldier, reveals another identity—a negotiator capable of learning restraint as deliberately as it learns aggression.

10.3 Immunity as Code

The immune system already computes. A T cell weighs evidence, integrates competing signals, and commits to an action that can be irreversible for the target cell. What has changed in the last decade is not the discovery that immunity makes decisions, but the confidence that those decisions can be *rewritten.*

In ordinary biology, a T cell's logic is the product of evolution and a lifetime of encounters. In synthetic biology, logic becomes a design material. Receptors can be swapped. Signaling pathways can be rerouted. Genes can be turned on only in the presence of a particular molecular combination. Cells can be engineered to behave less like generic soldiers and more like custom-built devices: living therapies that sense, decide, and act inside the body.

That ambition is easiest to grasp through a therapy that has already entered routine clinical practice: CAR-T cells.

A CAR, or chimeric antigen receptor, is a synthetic receptor installed on a T cell. "Chimeric" is a mythological word for a creature built from mismatched parts, and it is accurate here. A CAR typically combines an antibody-like recognition domain (the part that binds a target on a tumor) with intracellular signaling modules borrowed from T cell activation machinery. The result is a shortcut. Instead of relying on the usual pathway—tumor antigen chopped into peptides, presented on MHC, recognized by a T-cell receptor with the right specificity—a CAR-T cell can bind directly to a surface molecule on the tumor and interpret that binding as an order to attack.

If the immune system's natural recognition is like reading a sentence, CAR recognition is like scanning a QR code. It trades nuance for reliability.

One reason CAR-T therapy is so emotionally resonant is that it is made from the patient's own cells. The basic process is often described with a simplicity that feels almost science fictional: draw blood, isolate T cells, genetically modify them to express a CAR, expand them to large numbers, then infuse them back. A patient receives, in effect, an upgraded version of their own immunity.

The first FDA-approved CAR-T therapy, tisagenlecleucel (often known by the brand name Kymriah), was approved in 2017 for certain B-cell leukemias and lymphomas. The target was CD19, a molecule found on many B cells, including malignant ones. The choice of CD19 was strategic in a way that reveals how engineering thinks: a perfect tumor-specific target is rare, so clinicians accepted a trade. CD19 CAR-T cells would eliminate healthy B cells too, causing B-cell aplasia, but that condition can be managed with immunoglobulin replacement and careful monitoring. The therapy aimed for a controlled sacrifice: remove an entire cellular compartment that the body can partially compensate for, in order to remove a cancer that cannot be otherwise controlled.

The clinical stories that followed were often astonishing. Patients with advanced disease, sometimes after exhausting standard therapies, could enter deep remission. It would be dishonest to present this as universal; responses vary and relapses occur. Still, CAR-T therapy changed the emotional landscape of certain cancers. It also forced medicine to confront, in real time, what happens when a living therapy works *too* well.

The best-known failure mode is cytokine release syndrome (CRS). Cytokines are chemical messengers immune cells use to coordinate inflammation. If you suddenly infuse an army of cells designed to activate powerfully upon contact with a common tumor target, you can trigger a systemic storm of signals: fever, low blood pressure, organ strain, a patient who can deteriorate frighteningly fast. CRS is not a side effect in the way nausea might be a side effect; it is a direct expression of the therapy's potency.

The response to CRS is one of the clearest examples of immunology becoming systems engineering. Clinicians learned to anticipate the storm, grade its severity, and intervene with countermeasures. One key tool has been tocilizumab, an antibody that blocks the receptor for interleukin-6 (IL-6), a cytokine often central in the cascade. The existence of an "antidote" does not make the storm trivial, but it makes it governable, and governance is what allows a powerful technology to become standard care rather than a heroic last resort.

CAR-T therapy also highlights a second set of engineering problems: precision and control.

CD19 is, by design, not exclusive to tumor cells. Many solid tumors present an even harder challenge: they share markers with essential tissues, and they reside in environments that are metabolically harsh and immunologically suppressive. A CAR-T cell that attacks a solid tumor might also attack the healthy organ that shares the target antigen, and in a solid organ there may be no manageable equivalent of B-cell aplasia. "On-target, off-tumor" toxicity can be catastrophic.

This is where the idea of *immunity as code* becomes more than a metaphor. Engineers do not merely want stronger T cells. They want T cells that can implement conditional statements.

A familiar conditional statement in computing is an *AND* gate: execute the program only if condition A and condition B are both true. In immune terms, that might mean: kill only if two antigens are present, reducing the chance that a healthy tissue expressing only one antigen will be attacked. Another is a *NOT* gate: do not attack if a "do not harm" signal is present. These logical designs are attractive because biology is messy: no single marker perfectly distinguishes "tumor" from "self," but combinations can be more specific.

Synthetic biology has developed several ways to approximate such logic. One influential approach, introduced by Wendell Lim's lab in 2016, uses synthetic Notch receptors, often called synNotch. Natural Notch receptors are used in development to translate cell-to-cell contact into gene expression changes. synNotch repurposes this concept: binding of an external antigen can trigger the release of an engineered transcription factor inside the cell, which then turns on a chosen gene program. The critical feature is modularity. Recognition does not have to immediately trigger killing; it can trigger *preparation.*

A synNotch circuit can be used to implement an AND gate in a biologically intuitive way: antigen A triggers expression of a CAR against antigen B. In tissues where antigen A is absent, the CAR is not expressed, so antigen B alone cannot trigger killing. In tissues where antigen A is present, the CAR is produced, and then antigen B can activate the killing program. It is two-step verification, written in protein.

In later refinements, synNotch designs have also been used to localize powerful payloads to the right place. Instead of making a cytokine flood the entire body, a cell can be engineered to secrete an immune-stimulating molecule only when it detects a tumor-defining cue in its immediate neighborhood. The therapeutic action becomes spatially restricted: the drug is made where it is needed and remains minimal elsewhere. It is a strategy that treats toxicity not as an unfortunate accident but as a predictable consequence of distributing power too widely.

The underlying dream is simple to say and hard to achieve: keep the potency of immune activation while confining it to the right address.

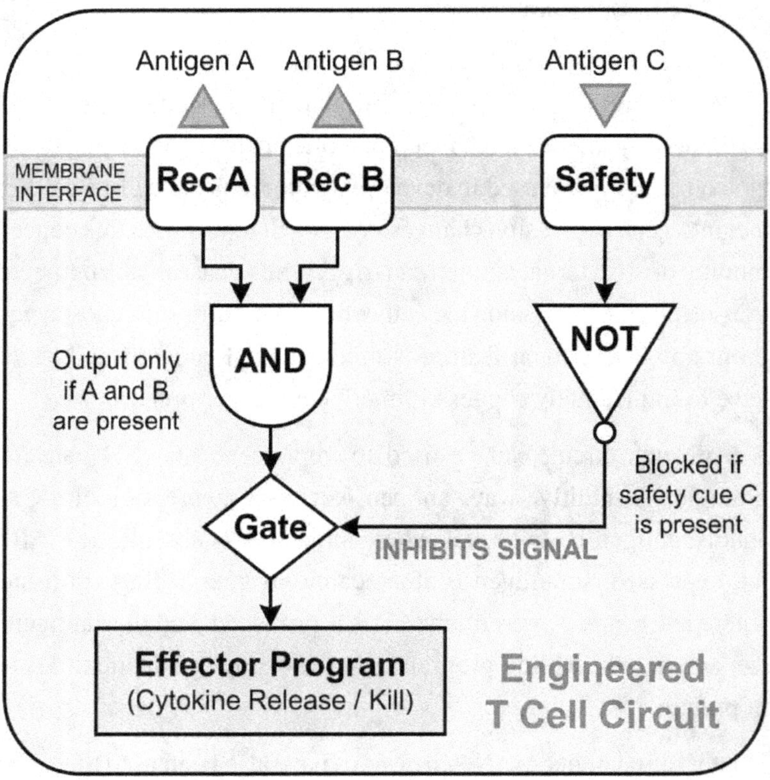

This schematic is deliberately cleaner than biology. Real engineered cells do not behave with the crisp certainty of digital circuits. Signals have thresholds, timings, and feedback loops. Receptors can be downregulated. Tumors can change what they express. Even so, the logic-gate metaphor has proven productive because it pushes designers to specify, explicitly, what a therapy should do in borderline cases.

Borderline cases are where the most important clinical decisions hide. A healthy tissue that expresses a low level of antigen B: should the cell ignore it? A tumor that expresses antigen A in patches: should the cell activate only locally? A blood vessel wall that expresses antigen C during inflammation: should that temporarily block anti-tumor activity, or should anti-tumor killing override the safety lock? Once these questions are asked, they demand answers not only in immunological terms but in ethical ones.

There is another layer of coding beneath receptors: editing the genome itself.

CRISPR, the gene-editing technology that moved from bacterial curiosity to laboratory staple with remarkable speed, has made it possible to modify human T cells in ways that would have sounded like science fiction in the era when CARs were first imagined. In a 2020 clinical report of CRISPR-engineered T cells in patients with refractory cancer, researchers performed multiplex edits, including disruption of endogenous T cell receptor components and disruption of *PDCD1*, the gene encoding PD-1. The goal was partly to prevent mispairing of introduced receptors with the cell's native receptor machinery, and partly to reduce inhibitory signaling that can limit function. The study showed feasibility and persistence of edited cells, while also underscoring the realities of risk: double-strand DNA breaks can lead to chromosomal rearrangements, a reminder that rewriting code inside living cells is not the same as editing text on a screen.

The larger aspiration is to make cellular therapies more consistent, more available, and safer. Autologous CAR-T manufacturing is personalized by necessity: each patient is their own factory. That is powerful, but also slow and expensive. It can fail when patients are too ill to provide enough healthy T cells. It encourages the development of "off-the-shelf" allogeneic products made from healthy donors, which in turn requires additional engineering to prevent graft-versus-host disease and rejection. Every improvement adds layers of code, and every layer introduces new debugging problems.

Yet the most striking aspect of this work is not technological bravado. It is a shift in what it means to understand immunity. For a long time, immunology was like meteorology: a science of complex patterns, often described in the language of correlation and probability. Synthetic biology turns parts of immunology into something closer to architecture. When you can build a functional circuit inside a cell, you are forced to specify the roles of each component and to confront the unspoken assumptions in the natural system.

There is also a human poignancy to the idea of immune cells as programmable devices. People living with cancer do not experience their disease as a molecular pathway; they experience it as scans, fatigue, decisions about work and family, hope and dread that can change from week to week. A therapy that recruits their own cells into a redesigned form of vigilance can feel like a restoration of agency. The body is no longer a passive battleground. It becomes an active participant, capable of being upgraded.

The challenge now is to make that upgrade wise. Power without discrimination leads to harm. Discrimination without power leads to failure. The most promising designs try to refuse the old bargain in which efficacy and safety are opposites. They aim for a different equation: build cells that can be strong *because* they are precise, and precise *because* they can decide, in context, when to hold back.

In that sense, the future of immune engineering may look less like building a better weapon and more like teaching a cell to follow rules so nuanced that they begin to resemble judgment.

www.ingramcontent.com/pod-product-compliance
Lightning Source LLC
LaVergne TN
LVHW022200200926
848481LV00006B/1792